Surfactants and Detergents - Updates and New Insights

Edited by Ashim Kumar Dutta

Published in London, United Kingdom

IntechOpen

Supporting open minds since 2005

Surfactants and Detergents - Updates and New Insights
http://dx.doi.org/10.5772/intechopen.95694
Edited by Ashim Kumar Dutta

Contributors
Raman Sureshkumar, Sk Janani, Dhanabal Palanisamy, Vai Yee Hon, Ismail B.M. Saaid, Rini Setiati, Muhammad Taufiq Fathaddin, Aqlyna Fatahanissa, Monia Renzi, Cristiana Guerranti, Serena Anselmi, Francesca Provenza, Andrea Blaskovic, Nur Liyana Ismail, Sara Shahruddin, Jofry Othman, Roger Saint-Fort, Gajendra Rajput, Niki Pandya, Darshan Soni, Harshal Vala, Jainik Modi, Saeeda Naqvi, Mohsen T.A. Qashqoosh, Faiza A. M. Alahdal, Yahiya Kadaf Manea, Swaleha Zubair

Notice
Statements and opinions expressed in the chapters are these of the individual contributors and not necessarily those of the editors or publisher. No responsibility is accepted for the accuracy of information contained in the published chapters. The publisher assumes no responsibility for any damage or injury to persons or property arising out of the use of any materials, instructions, methods or ideas contained in the book.

First published in London, United Kingdom, 2022 by IntechOpen
IntechOpen is the global imprint of INTECHOPEN LIMITED, registered in England and Wales, registration number: 11086078, 5 Princes Gate Court, London, SW7 2QJ, United Kingdom
Printed in Croatia

British Library Cataloguing-in-Publication Data
A catalogue record for this book is available from the British Library

Additional hard and PDF copies can be obtained from orders@intechopen.com

Surfactants and Detergents - Updates and New Insights
Edited by Ashim Kumar Dutta
p. cm.
Print ISBN 978-1-83962-896-2
Online ISBN 978-1-83962-897-9
eBook (PDF) ISBN 978-1-83962-898-6

Meet the editor

Dr. Ashim Kumar Dutta holds a Ph.D. in Physical Chemistry from Jadavpur University, India. He has worked extensively on the characterization of supramolecular assemblies at the air-water interface and on substrates using advanced spectroscopic and microscopic methods, namely, Förster resonance energy transfer (FRET), epifluorescence (EFM), atomic force microscopy (AFM), and near field scanning optical microscopy (NSOM), among others. He has thirty-six journal articles and nineteen patents to his credit. He has worked in Japan, Canada, and the United States under various international fellowships. Having started his career with Unilever in their home and personal care category, Dr. Dutta later moved into agrochemicals and worked for United Phosphorus Limited (UPL) and Indofil Industries Limited as their Head of Global Formulations. Presently, he is vice president (R&D) for India Glycols Limited. Dr. Dutta's research interests include novel surfactants, colloidal chemistry, structure-property correlation studies of supramolecular assemblies, and bio-mimetic and bio-inspired supramolecular systems.

Contents

Surfactants and Their Applications for Remediation of Hydrophobic Organic
Contaminants in Soils
by Roger Saint-Fort

Preface

Food, beverages, medicines, cosmetics, and detergents are ubiquitous in daily life. A close look at these diverse products reveals one common ingredient: surfactants. Because of its their wide and expansive applications, surfactants are a billion-dollar industry that is estimated to reach a market size of USD 57.8 billion by 2028, USD 24.8 billion of which will come from the market for natural surfactants. Although a large portion of the existing surfactant market is essentially synthetic, the availability of natural surfactants at the commercial scales and the growing need for them is directly related to the increasing consciousness of environmental change and the need to mitigate it. Natural surfactants derived from renewable resources like vegetable crop oils are considered environmentally "green and safe" because of their low eco-toxicity in addition to their highly biodegradable characteristics in contrast to most synthetic surfactants that have long degradation times, often several months, allowing the build-up of toxic waste in natural underwater reservoirs, rivers, and soil.

This book is a compilation of several chapters that highlight the use of specific surfactants and their functional properties in different applications that range from enhanced oil recovery (EOR) from ocean beds to nano-emulsion-based drug delivery systems for the treatment of cancer. Keeping in view a large audience consisting of experts as well as researchers new to the subject, we have avoided all mathematical derivations of formulas used across the book, but appropriate references have been provided for the reader interested in equations.

This volume consists of eight chapters presenting various perspectives and insights on both synthetic and natural surfactants used in different applications.

Chapter 1 examines the structure-property relationship connecting surface tension, torque, and bending rigidity of interfacial films to predict the micro-emulsion characteristics of an oil-surfactant-brine micro-emulsion representing an EOR system.

Chapter 2 studies the effects of sodium lignosulphonate (SLS) produced from sugar cane bagasse, which is a waste material, on the yield of oil recovered EOR processes as a function of SLS concentration and at different salinities. This study suggests the feasibility of profitably using bagasse that is usually burnt and that causes massive atmospheric pollution.

Chapter 3 compares the detergency of saponins extracted from tea leaves on fabrics with the detergency of a well-known commercial product. The results reveal comparable benefits and suggest the possibility of replacing a slowly degrading synthetic surfactant like lauryl alcohol benzyl sulphonic acid (LABSA) with saponins that are natural, green, and readily biodegradable.

Chapter 4 establishes the advantages of using microemulsion systems for drug delivery in the treatment of cancer. These systems are thermodynamically stable and the presence of lipids enhances the bioavailability of drugs resulting in quicker alleviation of symptoms.

Chapter 5 provides an overview of biosurfactants derived from renewable feedstock like vegetable oils and sugar and the advantages of using these industrially. It also discusses the various challenges in bulk production of these materials.

Chapter 6 examines the interaction of the antiulcer drug roxatidine acetate loaded on chitosan nanoparticles with serum albumin. It investigates the influence of a surfactant Tween 80 in the release of the active ingredient using fluorescence and circular dichroism spectroscopy.

Chapter 7 discusses the presence of microplastics and nanoparticles in water bodies and ecotoxicity arising therefrom. This study demonstrates that polyethylene and titanium dioxide nanoparticles largely enhance the ecotoxicity of water bodies endangering aquatic life.

Chapter 8 highlights the importance of surfactant-enhanced remediation of soil contaminants, especially hydrophobic organic contaminants.

I take this opportunity to thank all the authors for their excellent contributions and the staff at IntechOpen for their assistance throughout the publication process. I also wish to thank my wife Malabika and my two children Sourav and Siddharth for their help and cooperation. I dedicate this book to the loving memory of my parents.

Dr. Ashim Kumar Dutta
Vice President (R&D),
India Glycols Limited,
Uttarakhand, India

Chapter 1

Experimental and Computational Modeling of Microemulsion Phase Behavior

Vai Yee Hon and Ismail B.M. Saaid

Abstract

The phase behavior of microemulsions formed in a surfactant-brine-oil system for a chemical Enhanced Oil Recovery (EOR) application is complex and depends on a range of parameters. Phase behavior indicates a surfactant solubilization. Phase behavior tests are simple but time-consuming especially when it involves a wide range of surfactant choices at various concentrations. An efficient and insightful microemulsion formulation via computational simulation can complement phase behavior laboratory test. Computational simulation can predict various surfactant properties, including microemulsion phase behavior. Microemulsion phase behavior can be predicted predominantly using Quantitative Structure-Property Relationship (QSPR) model. QSPR models are empirical and limited to simple pure oil system. Its application domain is limited due to the model cannot be extrapolated beyond reference condition. Meanwhile, there are theoretical models based on physical chemistry of microemulsion that can predict microemulsion phase behavior. These models use microemulsion surface tension and torque concepts as well as with solution of bending rigidity of microemulsion interface with relation to surface solubilization and interface energy.

Keywords: surfactant, microemulsion, phase behavior, solubilization, chemical enhanced oil recovery, computational chemistry

1. Introduction

With growing global energy demand and depleting reserves, enhanced oil recovery (EOR) has become more important. Among various EOR processes, chemical EOR has been labeled an expensive process, with overwhelming parameters needed to describe a chemical EOR process and not practical to measure every one of them [1]. The chemical formulations for chemical EOR process consist of single or a combination of alkaline, surfactant and polymer. The traditional chemical EOR processes are polymer flooding, surfactant and alkaline flooding. Over the years, different modes of chemical flood injections were devised. There are the binary mix of alkali–surfactant (AS), surfactant-polymer (SP), alkaline-polymer (AP), and alkaline-surfactant-polymer (ASP) slug [2].

The research and development effort to design a robust chemical EOR formulation tailored to a specific field is challenging and laborious. To design a successful surfactant related chemical EOR formulation, Pope [3] highlighted 8 surfactant selection criteria. The surfactant must produce high solubilization ratio [i.e., low interfacial tension (IFT)] at optimum condition, commercialize at low cost, be feasible to be tailored to specific crude oil, temperature and salinity, comprise of highly branched hydrophobe with low adsorption onto reservoir rock, be insensitive to surfactant concentration above critical micelle concentration (CMC) and have minimal inclination to form liquid crystals, gels, macroemulsions and show rapid coalescence to microemulsion.

Surfactant solubilization can be determined via phase behavior laboratory test. The phase behavior of microemulsions formed in a surfactant-brine-oil system is complex. Phase behavior for a particular microemulsion system has been known to be measured experimentally [4]. The study of surfactant phase behavior consists of the determination of the number, composition, and structure of phases formed by surfactant systems at a given set of conditions (pressure, temperature, and system composition), in observance of Gibbs phase rule [5].

The relationship between solubilization and IFT is described commonly by Healy & Reed [6] correlations and the Chun-Huh [7] equation. Nevertheless, the latter equation is more commonly used. The term "solubilization" was introduced by McBain [8] to describe the increased solubility of a compound associated with the formation of micelles or inverted micelles. The mechanism to enhance solubility varies depending on the surfactant structure, the solvent type and the nature of the solubilized compound. The oil and water solubilization parameters, SP_o and SP_w respectively, are expressed in $SP_o = V_o/V_s$ and $SP_w = V_w/V_s$, where V_o, V_w, and V_s are the volumes of oil, water and surfactant contained within the micellar phase [9]. The solubilization ratio of water and oil phase is measured as either the volume of solubilized water (V_w) or oil (V_o) over volume of surfactant (V_s) in the microemulsion phase. The solubilization ratio of oil (V_o/V_s) increases with the increase in salinity, while the solubilization ratio of water (V_w/V_s) decreases with the increase in salinity [10]. The region in or near where the solubilization of oil and water intersects versus salinity is the optimal salinity.

2. Microemulsion and Winsor phase behavior

Microemulsions are dispersions of oil and water stabilized by surfactant molecules [11–13]. They can take on many structures such as water droplets in oil, oil droplets in water, sponge like, bicontinuous structures, and lamellar phase [14]. Unlike emulsions, they are thermodynamically stable. This because of the oil-water IFT is low enough (below 10^{-2} nM/M) to compensate the dispersion entropy. The interfacial energy is balanced by the dispersion entropy when the dispersion sizes are small enough, which is below 100 Å [15]. Emulsion has much larger dispersion sizes at approximately 1 μm. The different in size between microemulsion and emulsion explains their difference in properties and appearance; however, their fundamental difference is thermodynamic stability [9].

For a given set of conditions (temperature, composition), microemulsion displays well defined structures. The physical properties of microemulsion often undergo an abrupt change over a narrow concentration range [16]. It is generally accepted that this rapid change in the property over the range of concentration is due to the formation of surfactant aggregates or micelle in solution [9]. The nucleation of micelles is spontaneous, forming a structure that can vary between spherical and cylindrical depending on the surfactant molecular structure, solution composition and temperature. **Figure 1** depicted the intermicellar equilibrium and the associated phase changes. Micellar structure, S_1 is formed when the hydrophilic group of a surfactant are in contact with water while the hydrophobic group are

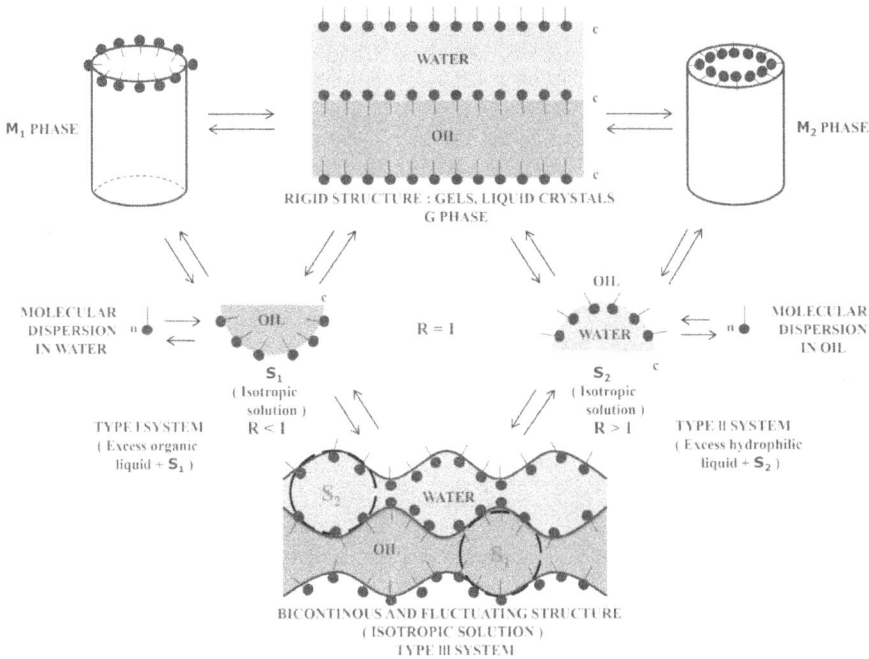

Figure 1.
Intermicellar equilibrium and associated phase changes [9].

gathered within the interiors of micelles to create small regions from which water is essentially excluded. When aggregates of surfactant form in apolar solvent, it is called inverted micelles, S_2. Inverted micelles promote the solubility of water in apolar solvents. Micellar aggregates have a finite mean lifetime, where their structures are mobile, the interfaces are flexible with a rapid exchange of molecules between the neighboring region and the aggregate [9]. Therefore, both S_1 and S_2 do form isotropic solutions with a bicontinuous and fluctuating structure in a given optimal condition. The microstructure of the sequence of states progressing from S_1 to S_2 is difficult to study and hence, these systems were large ignored until recent years, when it has been recognized that the isotopic solutions between S_1 and S_2 are agents which can significantly enhance oil recovery [9].

In less frequent cases, the microemulsion structure is made of elongated cylinders, eventually interconnected, or of distorted lamellar (sponge like), as depicted as M_1 and M_2 in **Figure 1**. These structures are encountered when the spontaneous curvature of the surfactant layer is small and approaches zero. On the contrary, sponge like structure can be significantly swollen by both oil and water. The G phase in **Figure 1** is liquid crystal with lamellar structure [15].

A microemulsion can exist in three types of systems. Winsor's introduced three types of simple phase diagram that are characterized by the nature of the polyphasic zone at low and moderate concentration. They are Winsor I (WI), Winsor II (WII) and Winsor III (WIII) [17]. Winsor I and II are also commonly known as II- and II+. Below a certain salinity, C_{seu}, the system is WI. Above a certain salinity, C_{sel}, the system is WII. If the salinity is between C_{seu} and C_{sel}, the system is WIII. In a WIII system, the IFT is lower than WI and WII [4]. WI and WII diagrams shows a characteristic of 2-phase behavior with water or oil microemulsion in equilibrium with the excess phase. WIII diagram shows a bicontinuous microemulsion in equilibrium with both the excess water and oil phases. WIII is considered to have the best probability of recovering additional oil. WII is considered to have the second-best chance to recover additional oil because it shows interaction between the aqueous phase and crude oil. Even though WI demonstrates interaction between the crude oil and the aqueous phase, it is considered to have poorer oil recovery potential than WII.

Winsor's phase behavior studies in the early 1950s introduced the R ratio of interactions of the adsorbed surfactant at the interface with the neighboring oil and water molecules as a criterion to take into account along with the effects of all formulation variables found in a ternary surfactant-oil-water (SOW) system, i.e., the surfactant head and tail characteristics, the nature of the oil, the aqueous phase salinity, as well as temperature and pressure [18].

In micellar solutions, 3 distinct regions can be identified: an aqueous region, W, an oil or organic region, O, and a surfactant region, C. The variation of the dispersing tendencies at the O and W faces of the C region is expressed qualitatively by Winsor [17] as:

$$R = \frac{(A_{CO} - A_{OO})}{(A_{CW} - A_{WW})} \tag{1}$$

where A_{CO} and A_{CW} are the interaction of surfactant molecules per unit area at the interface with oil and water respectively, A_{OO} is the interaction between two oil molecules, and A_{WW} is the interaction between two water molecules. Winsor described that an optimal microemulsion (Winsor III) is formed when the microstructure surface is flat, i.e., $R = 1$. When $R < 1$, there is a tendency to form oil-in-water emulsion (Winsor II or II+), whereas when $R > 1$, the tendency is to form water-in-oil emulsion (Winsor I or II-). Salager [18] presented a diagram (**Figure 2**) to link the Winsor R ratio with the observe phase type. $R < 1$, $R = 1$ and $R > 1$ correspond to WI (II-), WIII, and WII (II+) diagrams, respectively. This shows that

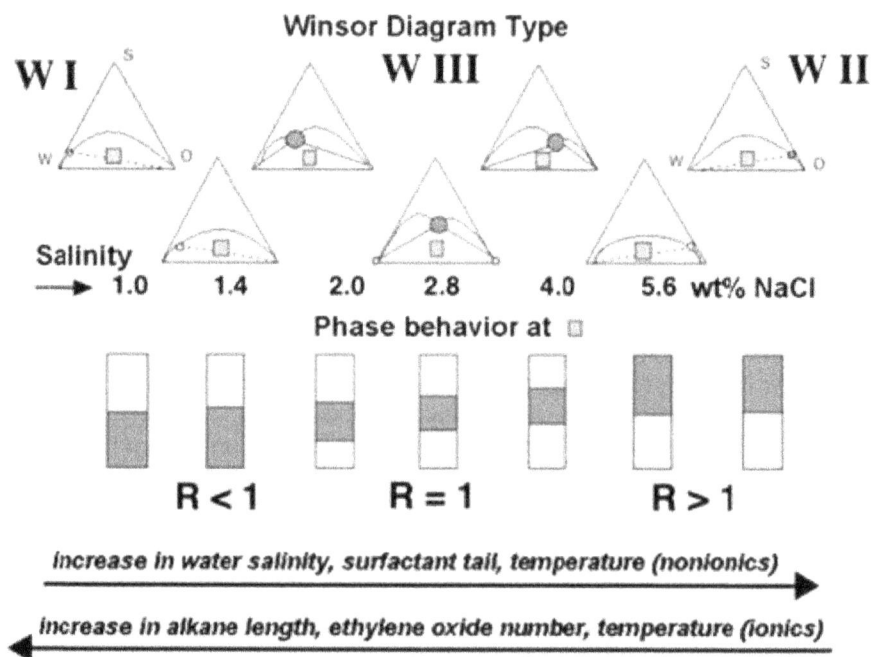

Figure 2.
Ternary phase diagram, test tube phase behavior and R ratio variations along a 1-dimensional formulation scan [18].

any formulation change that alters one of the interactions indicated in the ratio can increase or decrease R. When the formulation variation is properly selected to change R from $R < 1$ to $R > 1$ or vice versa, it changes the phase behavior from WI to WII or vice versa, with an intermediate WIII three-phase behavior at $R = 1$ [18].

3. Laboratory determination of phase behavior

Phase behavior laboratory test is conducted to assess surfactant performance in generating optimal microemulsion. The test involves the use of graduated cylinders with stopper and an explosion proof oven. A typical range of surfactant concentration and salinity for the phase behavior test are 0.25 to 2.5 wt.% and 5000 to 40,000 ppm respectively. The procedure starts with adding 50% of surfactant solution in brine into the graduated cylinder with 50% of crude oil of interest. The cylinder is then mixed vigorously and aged in the explosion oven for 14 days at reservoir temperature. The phase type and volume of the mixture are observed and measured on day-14 to identify the optimum microemulsion formulation and condition. The complete phase type descriptions are given in **Table 1**.

Type III is considered to have the best probability of recovering additional oil. Type II is considered to have the poorest chance to recover additional oil. Type II- is considered to have the second-best chance to recover additional oil because it shows interaction between the aqueous phase and crude oil. Even though Type II+ demonstrates interaction between the crude oil and the aqueous phase, it is considered to have poorer oil recovery potential than Type II-.

Figure 3 illustrates the relationships among phase behavior laboratory test observation, microemulsion structure and Winsor's R ratio.

4. Computational simulation of phase behavior

Apart from the commonly use of phase behavior laboratory test to assess microemulsion of a particular surfactant for chemical EOR application, it is viable to

Phase type	Phase type descriptions
II	Two fluid envelopes exist—A bottom aqueous phase and a top oil phase. No color is visible in the aqueous phase. The crude oil and aqueous phase volumes are equal to the volumes placed in the tube. The surfactant has been driven into the crude oil and no crude oil swelling has taken place (Type II+ phase behavior).
II-	Two fluid envelopes exist—A bottom aqueous phase and an oil phase. The bottom aqueous phase is colored due to surfactant carrying oil into the aqueous phase. The crude volume can be swollen due to the interaction with the surfactant (added and in-situ), but this is not a requirement for this designation.
III	Three or more fluid envelopes exist—A bottom aqueous phase, one or more middle emulsion phases, and a top crude oil phase. The aqueous phase can be colored with saponified acids (if alkali is presence) from the crude oil; however, this does not necessarily have to be the case.
II+	Two fluid envelopes exist—A bottom aqueous phase and a top crude oil phase. The bottom aqueous phase is clear because the surfactant (added and in-situ) resides in the crude oil phase. The crude oil phase is swollen due to surfactant carrying water into the crude oil phase.

Table 1.
Phase behavior type descriptions.

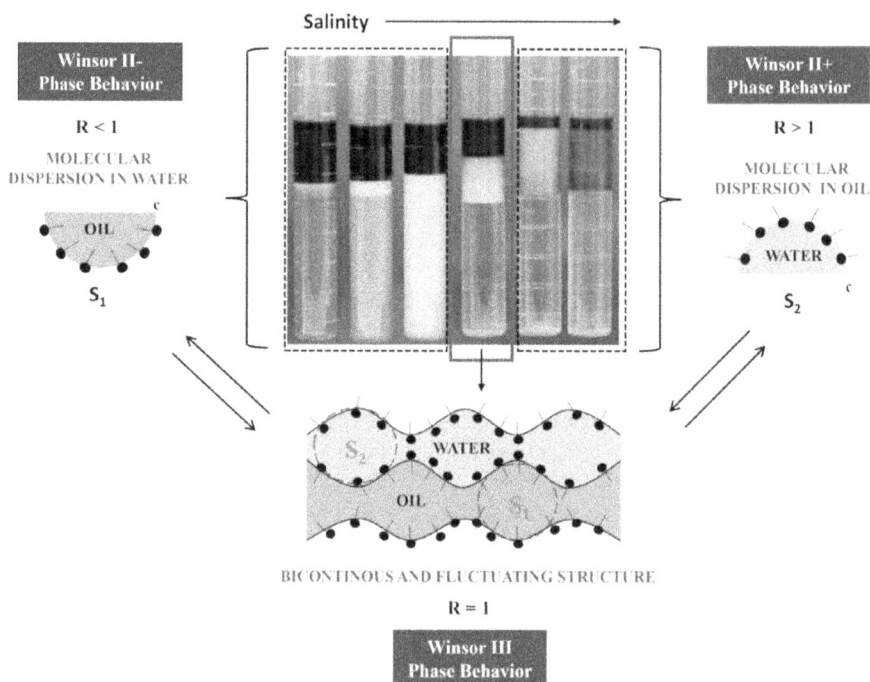

Figure 3.
Relationships among phase behavior laboratory test observation, microemulsion structure and Winsor's R ratio.

use computational simulation approach to predict surfactant performance. These computational simulation approaches include the use of various molecular modeling methods such as Monte Carlo (MC) simulations, Molecular Dynamics (MD), Dissipative Particle Dynamics (DPD) and upper scale modeling methods such as Quantitative Structure-Property Relationship (QSPR) approaches. Molecular modeling tools can be used to understand microscopic effects, predict surfactants' properties and finally to optimize structures and mixtures of surfactants [19]. Molecular modeling tools, in combination with recently developed intermolecular potentials, can provide precise information about microscopic phenomena and lead to accurate estimation of thermophysical properties [20–23]. Meanwhile, QSPR is an analytical method for breaking down a molecule into a series of numerical values describing its relevant chemical and physical properties. It remains as the focus of many studies aimed at the modeling and prediction of physicochemical and biological properties of molecules [24].

4.1 Quantitative structure-property relationship (QSPR)

QSPR is an approach to relate molecular descriptors with experimental values of properties based on statistical method. Its prediction accuracy is dependent on the size and quality of database and calculation of the relevant descriptors. There are many QSPR-like terms being used for more specific situations, such as Quantitative Structure-Activity Relationships (QSAR), Quantitative Structure-Toxicity Relationships (QSTR), Quantitative Property-Property Relationships (QPPR), Quantitative Sequence-Action Model (QSAM) and Quantitative Structure-Reactivity Relationships (QSRR) [25]. QSPR models have been developed to predict properties of pure surfactants only. Development of QSPR models for mixtures of surfactants is still a challenge [26].

Salager et al. [18] presented the use of Hydrophilic–Lipophilic Deviation (HLD) equations to attain optimum chemical EOR formulation for simple surfactant, oil and water system. The use of HLD concept to predict optimum surfactant formulation is a hybrid approach that combine HLD equations with experiments data, which is a QSPR approach. It was demonstrated that the phase behavior and optimum formulation can be manipulated with four main independent variables: brine salinity, oil alkane carbon number (ACN), surfactant parameter and temperature. However, these models are limited to simple system. Jin et al. [27] predicts the optimum surfactant salinity using HLD equation and measured parameters including the equivalent alkane carbon number (EACN), salinity, surfactant head dependent parameter, K_{surf} value and surfactant characteristic curvature, C_c. The work is extended to predict the IFT behavior using the Hydrophilic Lipophilic Difference-Net Average Curvature (HLD-NAC) equation of state.

Moreau et al. [28] applied the QSPR method to predict surfactant optimal salinity based on its correlation with surfactant structures. The QSPR models have been proven in reference conditions but they cannot be extrapolated to other conditions outside the application domain.

Budhathoki et al. [29] use the HLD equation to design the ratio of surfactant mixtures to form optimal microemulsion at reservoir condition. Correct surfactant head dependent parameter, K_{surf} and the surfactant temperature dependent parameter, αT values play a crucial role in the accuracy of the HLD method. Both K and αT values can be obtained via a combination use of HLD equation and a series of phase behavior laboratory work for each individual surfactant. This approach can reduce the experimental test matrix to design optimal surfactant mixture, but it is limited to surfactants with known K_{surf} and αT values through extensive phase behavior laboratory work.

4.2 Dissipative particle dynamics (DPD)

Many of the interesting phenomena that occur in complex fluids occur at the mesoscale, which is roughly defined as the spatio-temporal scales ranging from $10–10^4$ nm to $1–10^6$ ns [30]. These scales of simulation are not feasible using MD simulation. DPD is a coarse-grained type of molecular simulation technique which could reduce the length and timescale of molecular dynamics (MD) simulations, allowing simulation of large and complex system. DPD modeling method can reach large length scales by combining molecule groups into particles or beads, and long-time scales by replacing atomistic forces with soft effective forces [31]. DPD is widely popular simulation approach due to its algorithmic simplicity and huge versatility. By varying the conservative forces between coarse-grained beads, one can readily model complex fluids such as polymers, colloids, surfactants, membranes, vesicles and phase separating fluids [30]. DPD can give insights on spatial organization of surfactants, interesting mechanistic information for films evolution or trends on surface tensions regarding structure of the adsorbed tensioactive molecules at an interface [19]. However, the challenges for a successful DPD simulation is finding robust and general methods for parameterization of the simulation system [32–33]. This is an active research area with recent approach to apply machine learning for DPD parameterization [34].

A breakthrough approach by Fraaije et al. [35] demonstrated the use of surface torque analysis in simulating surfactant phase behavior with DPD, to determine the optimal brine salinity specifically. Prior to this, QSPR statistical approach have been the only known approach for decades in determining surfactant phase behavior. Buijse et al. [36] used Fraaije et al. approach to design EOR surfactant formulation by optimizing the surfactant head and tail composition as well as the use of co-solvent. Both Fraaije et al. and Buijse et al. work are applied in simple pure oil system. Further discussion on the work of Fraaije et al. is in Section 4.2.1.

Rekvig et al. [14] varies the surfactant chain length and topology to investigate the effect of surfactant structure and composition of the monolayer on the bending rigidity. This work of Rekvig et al. is of particular interest, where the linking of bending rigidity to surfactant structure in predicting the stability of microemulsions is demonstrated. This is important because it is agreed that the bending rigidity is a key parameter in understanding structure and phase behavior of microemulsion [37]. Further details of the approach by Rekvig et al. in determining bending rigidity is discussed further in the next Section 4.2.2.

4.2.1 Surface tension analysis

The work of Fraaije et al. [35] is the only found published work for direct determination of surfactant phase behavior with theoretical foundation based on physical chemistry of microemulsion. The calculation principal is based on an observation by Helfrich [38–39] that one can calculate the surface torque density (torque) from the first moment of the molecular stress profile, provided the surface is tensionless. A positive torque implies a tendency to bend toward the oil phase and form a microemulsion with oil droplets dispersed in an aqueous phase, while a negative torque a tendency to form a microemulsion that has water droplets dispersed in an oil phase. A surface with zero torque has an indifferent tendency where the system will form the optimal or balanced microemulsion with average zero curvature. Fraaije et al. run the DPD simulation including electrostatics and ion interactions with added salt, surfactants, and oil. The relationship between mechanical coefficients and the stress profile is expressed in Eq. (2) [39–40]:

$$M_n = \int \sigma(z) z^n dz \qquad (2)$$

where M_n is the stress moment, σ is the stress tensor and z is the coordinate perpendicular to the surface.

In mathematics, a moment of a function is a specific quantitative measure, used in both mechanics and statistics, of the shape of a set of points. For example, for a set of data points representing mass, the *0-th* moment is the total mass, the first moment divided by the total mass is the center of mass, and the second moment is the rotational inertia. Similarly, for a set of data points representing probability density, the zeroth moment is the total probability (i.e., one), the first moment is the mean, the second moment is the variance, and the third moment is the skewness. The *n-th* moment, M_n, of a real-valued continuous function $f(x)$ of a real variable about a value c is given is Eq. (3) below:

$$M_n = \int_{-\infty}^{\infty} (x - c)^n f(x) dx \qquad (3)$$

Fraaije et al. presented a surface tension analysis namely Method of Moments to shows that torque can be calculated based on Eq. (3). This is done by calculating the torque of an microemulsion interface from the first moment, M_1, only if the zeroth moment, M_0 (the interfacial tension) is zero exactly. Otherwise, both values of the neutral surface position and the interfacial tension has to be known. Meanwhile, in the tensionless limit, the value of z_s is inconsequential. The simulation is run across various salinities to find the optimal salinity at which the microemulsion surface torque is zero. Note that there is no clear-cut boundary on when a positive or negative torque transitions in to a small value (around zero). Therefore, the boundaries are somewhat gradual. Fraaije et al. demonstrated that the Method of Moments is principally correct by successfully deriving known empirical coefficients of decade-old QSPR models.

It was noted in Fraaije et al. work that the torque is related to the bending rigidity and spontaneous curvature through Eq. (4):

$$\tau \equiv kC_0 \tag{4}$$

where τ is the torque, k is the bending rigidity and C_0 is the spontaneous curvature. However, their approach does not allow for direct calculation of bending rigidity or the spontaneous curvature to deduce the stiffness of the interface. Furthermore, they have yet to attempt a full treatment of the compound mixture for application in actual crude oil system.

4.2.2 Interface fluctuation analysis

Microemulsion structure is governed by the elastic constants, the bending rigidity, k and the saddle splay modulus, k_s [41]. Bending rigidity characterizes the resistance of the interface toward bending. A low bending rigidity means large thermal undulations and low stability.

Rekvig et al. [14] used DPD to simulate surfactant monolayers on the interface between oil and water to calculate the bending rigidity by analyzing the undulation spectrum. The effect of surfactant density, chain length, adding co-surfactant and linear versus branched surfactant on bending rigidity are investigated. The results show that increase of the monolayer thickness has a larger effect on the bending rigidity than increasing the density of the layer. The bending rigidity also increases with surfactant chain length and is larger for linear than branched surfactants. Bending rigidity decrease linearly with mole fraction of short surfactants. Mixed film has a lower bending rigidity than the corresponding pure film for all mole fractions.

The work by Rekvig et al. [14] is reference with an earlier work by Goetz et al. [42] for lipid bilayer. Goetz et al. was the first to compute bending rigidity in molecular dynamics simulations. Rekvig et al. used a simple model of head, tail, water, and oil beads in DPD to capture the essential properties of ternary systems such as phase separation and adsorption. During the simulation at very low IFT, the interface is not strictly flat and undulatory waves can be observed (**Figure 4**).

These fluctuations of the interface are analyzed to compute bending rigidity. It is firstly done by characterizing the interface based on continuum theory [43]. This is then followed by the adoption of Helfrich [44] free energy of the interface [Eq. (5)]

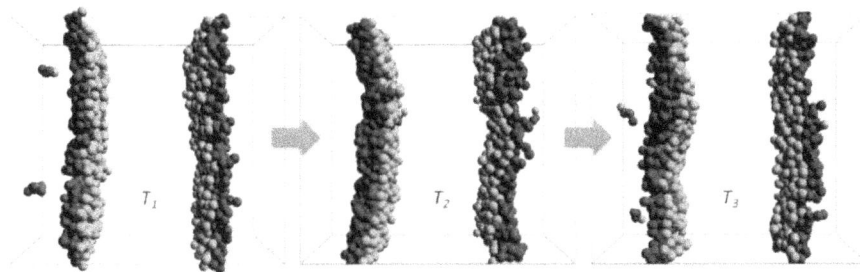

Figure 4.
Undulation wave of microemulsion interface over time (T1 to T3).

with local displacement from the average position of the interface to enable easier monitoring in simulations.

$$f_H = \gamma + \frac{1}{2}k(2H - c_0)^2 + k_s K \tag{5}$$

where H is the principal curvature of the surface at M (unique point), K is the gaussian curvature, depending on its sign, the surface will be curved like a sphere or like a saddle-splay, these number give us an idea of the surface's shape, k is the bending rigidity, k_s is the splay modulus which is related to the shape of the interface (K) and c_0 is the spontaneous curvature.

Bending rigidity is obtained by analyzing the undulation spectrum of the interface. Based on hypotheses from Rekvig et al. [14] using equipartition principal and fast fourier transform (FFT) application to decomposes the undulation signal into different wave lengths, Eq. (5) is transformed into Eq. (6) below:

$$\left\langle \left| \tilde{h}(q) \right|^2 \right\rangle = \frac{k_B T}{A} \left(\frac{1}{(\gamma q^2 + k q^4)} \right) \tag{6}$$

where $\tilde{h}$ is the approximation of local displacement from the average position of the interface, q is the wave vector, $\frac{2\pi}{\lambda}$, λ is the corresponding wavelength, k_B is Boltzmann constant (1.38×10^{23} J/K), T is absolute temperature in Kelvin, A is the interface area and γ is the interfacial tension.

Given the spectral intensity, S(q):

$$S(q) = \left\langle \left| \tilde{h}(q) \right|^2 A \right\rangle \tag{7}$$

Combining Eqs. (6) and (7), Eq. (8) is devised:

$$\frac{1}{S(q)q^2} = \frac{\gamma + k q^2}{k_B T} \tag{8}$$

Eq. (8) is used to fit the interface undulation spectrum analysis' results to estimate the bending rigidity, k. In DPD simulation, all units are normalized where $k_B T$ is unity.

Published experiment data on bending rigidity values for microemulsion system may be used as a reference for the model predicted bending rigidity values. However, such published data is scarce or not related to monolayers. Majority of the publications are generally focused on theory. Zvelindovsky [45] mentioned that the bending rigidity for a surfactant monolayer between water and oil is usually in the range of 1–20 kBT. Martínez et al. [46] performed MD simulation for SDS surfactant in dodecane and brine system at zero salinity. The associate bending rigidity is 1.3 kBT. SDS is sodium dodecyl sulfate or sodium lauryl sulfate, sometimes written as sodium lauril sulfate. Kegel et al. [47] found that the bending rigidity for SDS surfactant with alcohol in cyclohexane and brine system is around 1 kBT. Binks et al. [48] found a bending rigidity of around 1 kBT for AOT surfactant in nonane and a brine system at optimum salinity and surfactant concentration. AOT is a twin tailed, anionic surfactant with a sulfosuccinate head group stabilized as a salt by a sodium cation. It was also reported that the bending rigidity value depends on the alkane length, where bending rigidity decreases with increase in alkane length.

4.3 Monte Carlo (MC) and molecular dynamics (MD)

Both molecular modeling methods of MD and MC studies have been carried out for academic considerations, mainly on the surfactant aggregation process, instead of for industry application. This is mostly due to the use of atomistic description of the system that requires extensive computational power, which is not practical for industry application. As mention in Section 4.2.2, Goetz et al. [42] provides the first explicit connection between computer simulations with molecular resolution and elastic membrane models based on differential geometry. Goetz's method demonstrates a relationship between bending rigidity and the IFT.

5. Conclusion

Phase behavior of microemulsion is commonly assessed via laboratory study. These studies are straightforward but laborious especially when it involves a huge range of surfactant choices. Computational simulation is an alternative approach to provide insights into microemulsion phase behavior. There are limited computational simulation studies to predict surfactant phase behavior, whereby the widely used method since the beginning is empirical correlations as in QSPR approach. There are very few non-empirical approaches to predict surfactant phase behavior. These approaches are based on combination of physical chemistry of microemulsion surface tension, torque and bending rigidity concepts.

Acknowledgements

The authors would like to thank PETRONAS and University of Technology PETRONAS (Grant 015MD0-071) for permission and support to publish this book chapter.

Nomenclature

A	interface area
A_{CO}	interaction of surfactant molecules per unit area at the interface with oil
A_{CW}	interaction of surfactant molecules per unit area at the interface with water
A_{OO}	interaction between two oil molecules
A_{WW}	interaction between two water molecules
c	median of the excess stress profile
C	surfactant region
C_c	surfactant characteristic curvature
c_0	spontaneous curvature
C_{seu}	certain salinity 1
C_{sel}	certain salinity 2
f_H	free energy of the interface
G	liquid crystal with lamellar microemulsion structure
k	bending rigidity
k_B	Boltzmann constant (1.38×10^{-23} J·K^{-1})
k_b	bond strength
k_s	splay modulus
K_{surf}	surfactant head dependent parameter

K	gaussian curvature
M_n	n-th moment of a real-valued continuous function $f(x)$ of a real variable about a value c
M_1	interconnected elongated cylinder mircoemulsion structure
M_2	distorted lamellar elongated cylinder mircoemulsion structure
O	oil or organic region
q	wave vector
S_1	micellar structure
S_2	inverted micelles structure
SP_o	oil solubilization parameter
SP_w	water solubilization parameter
T	system temperature
V_o	volume of oil
V_w	volume of water
V_s	volume of surfactant
W	aqueous region
WI	Winsor I or II-
WII	Winsor II or II+
WIII	Winsor III
z	coordinate
γ	interfacial tension
τ	torque
σ	stress
λ	wavelength

Author details

Vai Yee Hon[1*] and Ismail B.M. Saaid[2]

1 PETRONAS, Kajang, Selangor, Malaysia

2 University of Technology PETRONAS, Seri Iskandar, Perak, Malaysia

*Address all correspondence to: honvaiyee@petronas.com.my

IntechOpen

References

[1] Sheng JJ. Modern Chemical Enhanced Oil Recovery: Theory and Practice. United States: Gulf Professional Publishing; 2011

[2] Gbadamosi AO, Junin R, Manan MA, Agi A, Yusuff AS. An Overview of Chemical Enhanced Oil Recovery: Recent Advances and Prospects. Vol. 9. Berlin Heidelberg: Springer; 2019

[3] Pope GA. Overview of chemical EOR center for petroleum and geosystems engineering. In: Casper EOR Workshop. Austin, TX: The University of Texas at Austin; 2007

[4] Sheng JJ. Modern Chemical Enhanced Oil Recovery. Amsterdam, Netherlands: Elsevier Inc.; 2011

[5] Marques EF, Silva BFB. Surfactants, phase behavior. In: Tadros T, editor. Encyclopedia of Colloid and Interface Science. Berlin, Heidelberg: Springer; 2013. pp. 1290-1333

[6] Healy RN, Reed RL. Immiscible microemulsion flooding. SPE Journal. 1977;**17**:129-139

[7] Huh C. Interfacial tensions and solubilizing ability of a microemulsion phase that coexists with oil and brine. Journal of Colloid and Interface Science. 1979;**71**(2):408-426. DOI: 10.1016/0021-9797(79)90249-2

[8] McBain JW. Advances in Colloid Science 1. Geneva: Interscience Publishers Inc; 1942

[9] Bourrel M, Schechter RS. Microemulsions and Related Systems. New York: Marcel Dekker; 1998

[10] Elraies KA, Ahmed S. Mechanism of surfactant in microemulsion phase behavior. Journal of Applied Sciences. 2014;**14**:1049-1054

[11] Overbeek JTG. The first rideal lecture. Microemulsions, a field at the border between lyophobic and lyophilic colloids. Faraday Discussions of the Chemical Society. 1978;**65**:7

[12] Overbeek JTG. Microemulsions. Proceedings. 1986;**B89**:61

[13] Gennes D, Taupln C. Microemulsions and the flexibility of oil/water interfaces. Journal of Physical Chemistry. 1982;**86**(13):2294-2304. DOI: 10.1021/j100210a011

[14] Rekvig L, Hafskjold B, Smit B. Simulating the effect of surfactant structure on bending moduli of monolayers. The Journal of Chemical Physics. 2004;**120**(10):4897-4905. DOI: 10.1063/1.1645509

[15] Langevin D. Microemulsions—Interfacial aspects. Advances in Colloid and Interface Science. 1991;**34**: 583-595. DOI: 10.1016/0001-8686(91)80059-S

[16] Preston WC. Some correlating principles of detergent action. The Journal of Physical and Colloid Chemistry. 1948;**52**:84

[17] Winsor PA. Binary and multicomponent solutions of amphiphilic compounds. Solubilization and the formation, structure, and theoretical significance of liquid crystalline solutions. Chemical Reviews. 1968;**68**(1). DOI: 10.1021/cr60251a001

[18] Salager JL, Forgiarini AM, Márquez L, Manchego L, Bullón J. How to attain an ultralow interfacial tension and a three-phase behavior with a surfactant formulation for enhanced oil recovery: A review. Part 2. Performance improvement trends from Winsor's premise to currently proposed inter- and intra-molecular mixtu. Journal of Surfactants and Detergents. 2013;**16**(5):

631-663. DOI: 10.1007/s11743-013-1485-x

[19] Creton B, Nieto-Draghi C, Pannacci N. Prediction of surfactants' properties using multiscale molecular modeling tools: A review. Oil and Gas Science and Technology. 2012;**67**(6): 969-982. DOI: 10.2516/ogst/2012040

[20] Gubbins KE, Moore JD. Molecular modeling of matter: Impact and prospects in engineering. Industrial and Engineering Chemistry Research. 2010; **49**(7):3026-3046. DOI: 10.1021/ie901909c

[21] Theodorou DN. Progress and outlook in Monte Carlo simulations. Industrial and Engineering Chemistry Research. 2010;**49**(7):3047-3058. DOI: 10.1021/ie9019006

[22] Maginn EJ, Elliott JR. Historical perspective and current outlook for molecular dynamics as a chemical engineering tool. Industrial and Engineering Chemistry Research. 2010; **49**(7):3059-3078. DOI: 10.1021/ie901898k

[23] Moeendarbary E, Ng TY, Zangeneh M. Dissipative particle dynamics: Introduction, methodology and complex fluid applications—A review. International Journal of Applied Mechanics. 2009;**1**(4):737-763. DOI: 10.1142/S1758825109000381

[24] Liu F, Cao C, Cheng B. A quantitative structure-property relationship (QSPR) study of aliphatic alcohols by the method of dividing the molecular structure into substructure. International Journal of Molecular Sciences. 2011;**12**:2488-2462

[25] Díaz HG. Bioinformatics and quantitative structure-property relationship (QSPR) models. Current Bioinformatics. 2013;**8**(4):387-389

[26] Muratov EN, Varlamova EV, Artemenko AG, Polishchuk PG,

Kuz'Min VE. Existing and developing approaches for QSAR analysis of mixtures. Molecular Informatics. 2012; **31**(3–4):202-221. DOI: 10.1002/minf.201100129

[27] Jin L et al. Development of a chemical flood simulator based on predictive HLD-NAC equation of state for surfactant. SPE Annual Technical Conference and Exhibition; 2016. p. 20. DOI: 10.2118/181655-ms

[28] Moreau P, Maldonado A, Oukhemanou F, Creton B. Application of quantitative structure-property relationship (QSPR) method for chemical EOR. In: Proceedings of 17th European Symposium on Improved Oil Recovery. OnePetro; 2013. DOI: 10.2118/164091-MS

[29] Budhathoki M, Hsu TP, Lohateeraparp P, Roberts BL, Shiau BJ, Harwell JH. Design of an optimal middle phase microemulsion for ultra high saline brine using hydrophilic lipophilic deviation (HLD) method. Colloids Surfaces A Physicochemical and Engineering Aspects. 2016;**488**:36-45. DOI: 10.1016/j.colsurfa.2015.09.066

[30] Español P, Warren PB. Perspective: Dissipative particle dynamics. The Journal of Chemical Physics. 2017;**146**: 150901

[31] Gao L, Shillcock J, Lipowsky R. Improved dissipative particle dynamics simulations of lipid bilayers. The Journal of Chemical Physics. 2007;**126**(1):1-8. DOI: 10.1063/1.2424698

[32] Buijse MA, Tandon K, Jain S, Handgraaf JW, Fraaije J. Surfactant optimization for eor using advanced chemical computational methods. In: SPE Improved Oil Recovery Symposium. Tulsa, Oklahoma: Society of Petroleum Engineers; 2012. p. 12. DOI: 10.2118/154212-MS

[33] Khedr A, Striolo A. DPD parameters estimation for simultaneously

simulating water-oil interfaces and aqueous non-ionic surfactants Abeer Khedr and Alberto Striolo* chemical engineering department, University College London, United Kingdom. Journal of Chemical Theory and Computation. 2018;**14**:6460-6471

[34] Mcdonagh JL, Shkurti A, Bray DJ, Anderson RL, Pyzer-Knapp EO. Utilizing machine learning for efficient parameterization of coarse grained molecular force fields. Journal of Chemical Information and Modeling. 2019;**59**(10):4278-4288. DOI: 10.1021/acs.jcim.9b00646

[35] Fraaije JGEM, Tandon K, Jain S, Handgraaf JW, Buijse M. Method of moments for computational microemulsion analysis and prediction in tertiary oil recovery. Langmuir. 2013;**29**(7):2136-2151. DOI: 10.1021/la304505u

[36] Buijse M, Tandon K, Jain S, Jain A, Handgraaf JW, Fraaije JGEM. Accelerated surfactant selection for EOR using computational methods. In: SPE Enhanced Oil Recovery Conference. 2013. pp. 558-567. DOI: 10.2118/165268-ms

[37] Gradzielski M. Bending constants of surfactant layers. Current Opinion in Colloid & Interface Science. 1998;**3**(5):478-484. DOI: 10.1016/S1359-0294(98)80021-6

[38] W. Helfrich. Amphiphilic Mesophase Made of Defects. Physics of Defects. 1981; **35**:716-755

[39] Helfrich W. Elasticity and thermal undulations of fluid films of amphiphilies. Liquids at Interfaces. 1990;**48**:212-237

[40] Szleifer I, Kramer D, Ben-Shaul A, Gelbart W, Safran S. Molecular theory of curvature elasticity in surfactant films. The Journal of Chemical Physics. 1990;**92**:6800-6817

[41] Holderer O, Frielinghaus H, Monkenbusch M, Klostermann M, Sottmann T, Richter D. Experimental determination of bending rigidity and saddle splay modulus in bicontinuous microemulsions. Soft Matter. 2013;**9**(7):2308-2313. DOI: 10.1039/c2sm27449c

[42] Goetz R, Gompper G, Lipowsky R. Mobility and elasticity of self-assembled membranes. Physical Review Letters. 1999;**82**(1):221-224. DOI: 10.1103/PhysRevLett.82.221

[43] Safran SA. Statistical Thermodynamics of Surfaces, Interfaces, and Membranes. 1st ed. Boulder, CO: Westview Press; 2003

[44] Helfrich W. Elastic properties of lipid bilayers: Theory and possible experiments. Zeitschrift für Naturforschung. 1973;**28c**:693-703

[45] Zvelindovsky AV. Nanostructured Soft Matter: Experiments, Theory, Simulation and Perspectives. Berlin, Germany: Springer; 2007

[46] Martínez H, Chacón E, Tarazona P, Bresme F. The intrinsic interfacial structure of ionic surfactant monolayers at water-oil and water-vapour interfaces. Proceedings of the Royal Society A: Mathematical, Physical and Engineering Sciences. 2011;**467**(2131):1939-1958. DOI: 10.1098/rspa.2010.0516

[47] Kegel WK, Bodnàr I, Lekkerkerker HNW. Bending elastic moduli of the surfactant film and properties of a Winsor II microemulsion system. The Journal of Physical Chemistry. 1995;**99**(10):3272-3281. DOI: 10.1021/j100010a043

[48] Binks BP, Kellay H, Meunier J. Effects of alkane chain length on the bending elasticity constant k of AOT monolayers at the planar oil-water interface. EPL. 1991;**16**(1):53-58. DOI: 10.1209/0295-5075/16/1/010

The Importance of Microemulsion for the Surfactant Injection Process in Enhanced Oil Recovery

Rini Setiati, Muhammad Taufiq Fathaddin and Aqlyna Fatahanissa

Abstract

Microemulsion is the main parameter that determines the performance of a surfactant injection system. According to Myers, there are four main mechanisms in the enhanced oil recovery (EOR) surfactant injection process, namely interface tension between oil and surfactant, emulsification, decreased interfacial tension and wettability. In the EOR process, the three-phase regions can be classified as type I, upper-phase emulsion, type II, lower-phase emulsion and type III, middle-phase microemulsion. In the middle-phase emulsion, some of the surfactant grains blend with part of the oil phase so that the interfacial tension in the area is reduced. The decrease in interface tension results in the oil being more mobile to produce. Thus, microemulsion is an important parameter in the enhanced oil recovery process.

Keywords: enhanced oil recovery, emulsion, middle-phase emulsion, microemulsion, surfactant

1. Introduction

The general oil production activity is categorized into three phases: primary, secondary and tertiary. The primary recovery is limited to activities that naturally cause hydrocarbon to rise toward the surface with artificial technology, such as a submersible pump. Secondary recovery is conducted by utilizing water and gas injections that fill lower layers of oil and drive them into the surface. In secondary recovery, the refineries are often found unable to produce although they possess relatively large saturation levels between 30% and 40% of their initial reserves. These oils are trapped inside the reservoir and are unable to move. This situation often happens because oil grains rest on the surface of rocks and produce high interface tension levels. Enhanced oil recovery (EOR) chemical injection is aimed at freeing these trapped oil grains in the reservoir. EOR is a tool to produce oil in the third phase, also known as tertiary recovery effort. At this stage, a lot of effort is made to improve oil production after primary recovery and secondary recovery efforts. The way to further increase oil production is through the tertiary recovery method or EOR. Although more expensive to employ on a field, EOR can increase production from a well to up to 75% recovery.

EOR is the main concentration of petroleum engineering that is related to higher level of oil recovery based on physics principles through different techniques utilization. A number of formulas are utilized to improve oil recovery that involves important parameters such as injection rate, oil production rate, movement efficiency, mobility, and reservoir characteristics and fluids [1]. EOR process is determined to enhance the ability of oil to move toward the well by injecting water, chemical compound or gas into the reservoir that would alter oil's physical nature. Its main goal is to produce optimum amount of oil after primary and secondary productions are implemented. EOR process includes thermal, chemical and miscible methods. EOR process is designed to recover the remaining amount of oil after primary and secondary recovery by enhancing oil movement and volumetric sweeping efficiency [2].

Chemical EOR is an efficient oil recovery technique to improve oil flow and to produce oil trapped in the reservoir. This EOR method is implemented by using chemical injection to improve oil recovery levels [3]. The implementation of this method is by inserting a long molecular chain, such as polymer or surfactant, through injection into the reservoir in order to improve waterflooding efficiency or to improve surfactant effectiveness, which acts as a sweeper that lowers interface tension level that prevents oil from flowing toward the reservoir.

EOR process that involves injecting into the reservoir is aimed at forming middle-phase emulsion in order to lower interfacial tension level. This condition will allow oil grains to move easily and to flow into production wells to be produced.

2. Methodology

Observation on mixed-phase behavior of reservoir fluid and injection fluid can be categorized as lower-phase emulsion, microemulsion (middle-phase emulsion), upper-phase emulsion and sediment. Mixed phase that forms microemulsion represents miscible displacement condition. Meanwhile, mixed phase that forms upper or lower phase represents immiscible displacement condition. Emulsion is a system of two phases, where one of the liquids is dispersed in another liquid in the form of small droplets [4].

Emulsion can be defined as droplet dispersion from one liquid to another one that is immiscible. These droplets, also known as the dispersed phase, drop into a second liquid called the continuous phase. To create a stable emulsion, the droplets should be kept in dispersed condition by adding surfactant or co-surfactant [5, 6]. Generally, in terms of the phase, two microemulsion types are often produced, namely water in oil (w/o) and oil in water (o/w). The two types of microemulsions imply that w/o emulsion describes that water acts as the dispersed phase, while w/o emulsion clearly implies that oil is the dispersed phase. Based on the size of the dispersed droplets, microemulsion and macroemulsion can be specifically identified, 5–100 nm for microemulsion and >100 nm for macroemulsion. The other difference is that microemulsions are often categorized as a thermodynamically stable system, while macroemulsions are the kinetically stable ones. The colors of the two types are also different, transparent with low viscosity on microemulsions and opaque with high viscosity on macroemulsions. The unique characteristics make the utilization of these two types of emulsions a vital part of the study.

Emulsions from immiscible mixtures are often results of stable and even ones. Salinity, temperature, oil type and surfactant used are the most determining factors that drive the mixture to become stable. There are two types of emulsions based on the type of liquid that serves as an internal or external phase, namely:

a. Emulsion type O/W (oil in water) is an emulsion consisting of oil droplets dispersed into water, oil as the internal phase and water as the external phase.

b. Emulsion type W/O (water in oil) is an emulsion consisting of water droplets dispersed into the oil, water as the internal phase and oil as the external phase.

According to [7] microemulsion phase, behavior is very much complex and highly depends on several determining parameters, which include the type and concentration of surfactant, co-solvent, hydrocarbon, salinity of salt water, temperature and low level of pressure. There is no formula to describe simple microemulsion. Based on that, phase behavior on a microemulsion system must be experimentally observed. Generally, to present microemulsion-phase behavior, ternary diagram is used as the main tool, although the pressure and temperature are considered at a constant level. Ternary diagram is a highly benefitting tool due to its ability to simultaneously and wholly represent phase composition and the relative numbers. **Figure 1** shows the schemes of ternary diagram. On ternary diagram, water, oil and surfactant are represented on a triangle-shaped model, with each concentration level represented on units of molar, mass or volume. The system can exist in a phase where the solubility ratio of water and oil is high. **Figure 1** shows ternary diagram of microemulsion system [9]. **Figure 1** also shows that there are two areas where oil can be recovered; in single-phase area that is dominantly filled with oil and in a number of parts or multiphase that includes a mixture of a number of phases, oil, water and surfactant.

Surfactant is a molecule that usually contains both head and tail poles [10]. Surfactant molecule arranges itself based on a number of intermolecular and intramolecular forces. For example, when surfactant is mixed with oil and water, surfactant is formed on oil and water interface, which leads to a thermodynamically benefitting condition [11]. Oil in water emulsion can happen in two ways: where oil molecule is incorporated in a micelle or in between surfactant tail, or it can also be incorporated in the hydrophobic core, as illustrated in **Figure 2**.

Microemulsion (multiphase part) is divided into three categories, namely lower phase (l), upper phase (u) and middle phase (m), in equilibrium with oil phase, water phase or both. Microemulsion, as observed by Winsor [13], shows the phenomenon when water (or brine) and organic compound (both mixed and single compounds) can be mixed with a suitable number on surfactant equilibrium system. There are four main types of equilibrium systems: type I, II, III and IV, as shown in the following **Figure 3**.

Figure 3 shows that Winsor I is a dominant emulsion formed in the water phase, where surfactant is mostly mixed with water. Winsor II is an emulsion formed of oil phase, where surfactant is mostly mixed with oil. Meanwhile, Winsor III is a

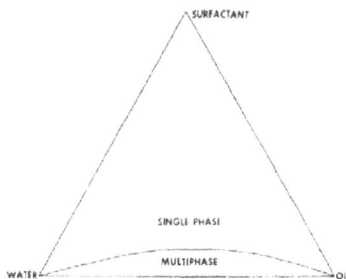

Figure 1.
Ternary diagram of micro-emulsion system [8].

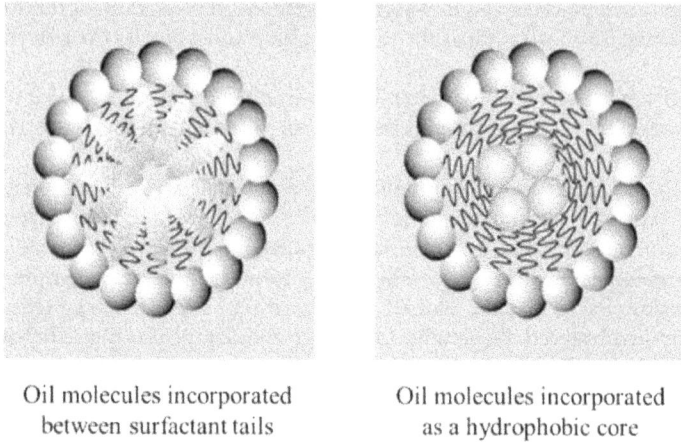

Oil molecules incorporated
between surfactant tails

Oil molecules incorporated
as a hydrophobic core

Figure 2.
Surfactant molecule and oil molecule [12].

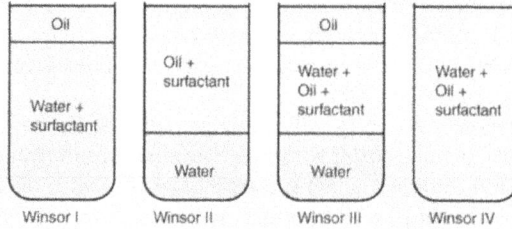

Figure 3.
Development of microemulsion [14].

middle-phase emulsion where surfactant is mixed with both oil and water. Winsor III is the expected result on surfactant injection mechanism. Microemulsion phase formed during this level of balance is known as the middle-phase emulsion. In this condition, organic and water phases are located in the phase that holds all three components [13, 14]. In Winsor I and II, the tie horizontal line should be at a similar level. Behavior formed by Winsor III can be seen in **Figure 4**; the polyphonic area includes three zones that are surrounded by the other two zones. Meanwhile, the

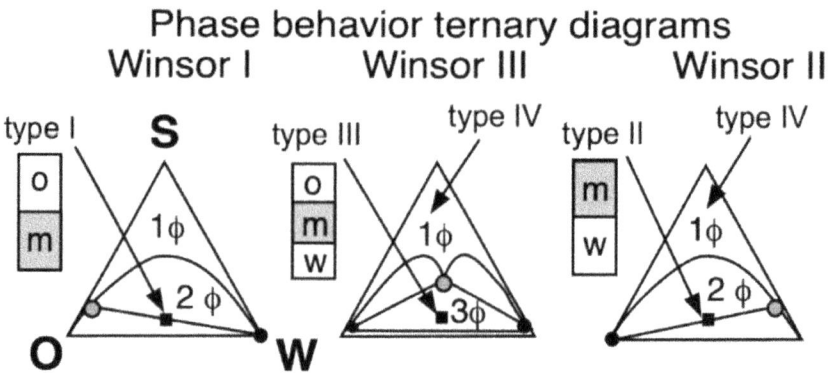

Figure 4.
Winsor I, II, III diagrams [15].

composition system of the phase is formed in three different phase zones, namely surfactant middle phase emulsion, and the other two phases, which basically consist of liquid and oil. This phase is also known as the middle phase because it is formed between oil and water due to its medium-density level. These three-phase systems are extensively studied in order to achieve better oil recovery level [15]. Based on that, Winsor III emulsion is preferred to improve oil recovery. The microemulsion formation mechanism of the three emulsion types of equilibrium phases can be seen in the following figures.

Single-phase area is located in the zone with high level of surfactant concentration, the three-phase area in the middle zone, and the lobe of two phases are located at the upper right and left areas of the triangle. There is a third two-phase area located on the lower surfactant under the three-phase area. Winsor III ternary phase behavior itself is included in areas close to saltwater/oil axis limited by the triangle. Winsor III is also known as Winsor type III.

Between type I and II, a horizontal tie line with similar surfactant condition between the two phases is expected. The type III phase behavior can be seen in **Figure 2**. The polyphonic area includes three types of zones, which are surrounded by two-phase zones. The composition system is actually located in the three-phase zones, which are categorized into surfactant-rich area (shaded area), oil and water. This phase is called as middle phase due to its location, which is in the middle of oil and water. Due to its accurate identification, the system is studied in an extensive manner in order to achieve better oil recovery level [1]. Based on that, to improve oil recovery level, type III emulsion is preferred (**Figure 5**). The type of O/W emulsion shown in **Figure 5** can be seen in general system microemulsion in **Figure 6**, where this microemulsion will release oil trapped in the rock grains as shown in **Figure 7**.

In microemulsion system, there is a rapid exchange of individual components between different types of environments and also a consistently fluctuated interface film. In microemulsion, there are two main phases that highly depend on its composition, which are the droplet phase and the bi-continual phase [19]. Droplet phase occurs in high concentration water; microemulsion contains oil droplets that are covered by the interface layer, which consists of co-surfactant part and freely dispersed surfactant in continuous phase, which is the water phase and forms o/w microemulsion. However, when water concentration is limited or lower, reverse situation occurs where water droplets get dispersed in oil and form w/o

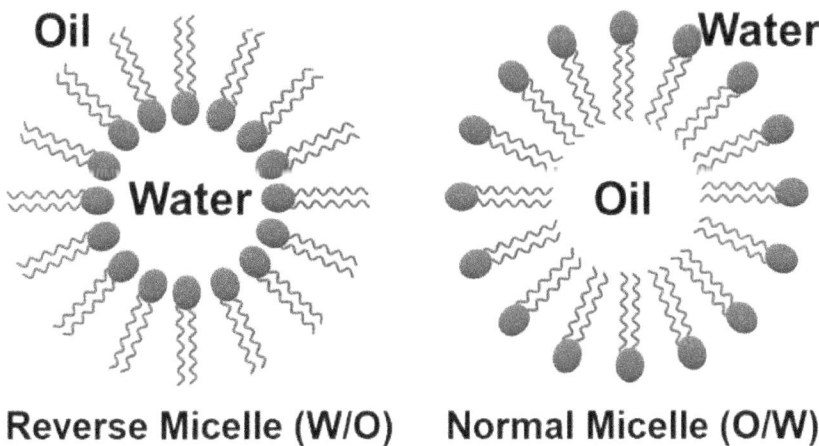

Figure 5.
Microemulsion formation mechanisms [16].

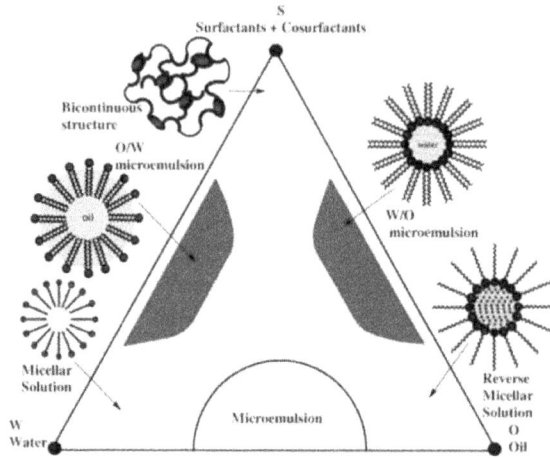

Figure 6.
Microemulsion systems of oil (o), water (w) and surfactant + co-surfactant (S) [17].

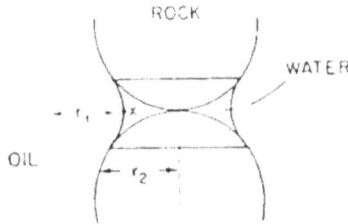

Figure 7.
Water trapped between two grains of sand on water-wet reservoir [18].

microemulsion [20]. Meanwhile for bi-continual phase, the transition is more stable than o/w to w/o microemulsion alteration system by changing the amounts of oil and water. The area formed in the middle possesses more or less similar fraction of water and oil. Generally, it contains bi-continual or thin structures where the two phases continually fluctuate until they almost reach zero [21, 22].

With the surfactants, water and oil mixed, emulsification, which can reduce the water-oil interfacial tension, occurs, so the capillary pressure in the pore narrowing area decreases. Because the capillary pressure is reduced, the resistance to oil flow will also be reduced and the oil will move more easily to be produced. Capillary pressure (P_c) is defined as the difference in pressure that exists between the surfaces of two immiscible fluids (liquid-liquid or liquid-gas) as a result of the interface between them. The pressure difference between these two fluids is the pressure difference between the "non-wetting phase" (P_{nw}) fluid and the "wetting phase" (P_w) fluid. In general, reservoirs are water-wet, so water tends to stick to the rock surface, while oil lies between the water phases. The character of oil-wet in reservoir rock conditions is not expected to occur because it will cause the amount of oil left in the reservoir rock when produced to be greater than water-wet. A liquid is said to wet a solid if the adhesion stress is positive ($\theta < 90°$), which means the rock is water-wet, whereas if the water does not wet the solid, then the adhesion stress is negative ($\theta > 90°$), meaning the rock is oil-wet. This wettability has an important role in the behavior of the reservoir because it will cause capillary pressure, which will provide an impetus so that oil or gas can move. In the reservoir, water is usually the wetting phase, while oil and gas are the non-wetting or non-wetting phases.

Capillary pressure in pored rocks depends on the size of pores and the type of fluid. Quantitatively, the relationship can be formulated as follows:

$$P_c = P_o - P_w = \sigma \left(\frac{1}{r_1} + \frac{1}{r_2} \right) \tag{1}$$

P_c = capillary pressure (pressure unit)
σ = interfacial tension
r_1 & r_2 = curve radius on interface area
P_o & P_w = pressure on oil and water phases

Capillary number theory is considered to be the basic theory in chemical injection in the EOR process. The basic mechanism of chemical flooding in EOR can be summarized into mobility control based on increasing sweep efficiency and based on capillary number theory, which increases displacement efficiency. In surfactant injection, changes in wettability and imbibition will occur in some cores that have heterogeneity at the pore scale. The heterogeneity in question is the difference in wettability; if the difference is greater, the imbibition will be more visible [23]. These factors affect the displacement efficiency (E_d), in addition to the capillary number (N_c) and the emulsion effect that occurs when surfactant injection. Emulsion will change the behavior of the permeability and wettability phase so that imbibition will occur.

For the purposes of exploration and exploitation of petroleum, water-wet formations are easier to perform oil recovery. This is because the water-wet formation has a high oil saturation, which means that water wets and fills the smallest pores in the rock grains and the oil is above the water so that it is easier for oil to move and spill because there is no adhesion force to the surface of the rock. This situation allows the increase in oil recovery in formations with sandstone reservoirs by injection of water into the formation so that it can increase the volume of water and encourage oil to come out and be produced. Carbonate reservoirs, which are oil-wet, require EOR by chemical injection with the aim of changing the chemical structure of the rock surface so that oil no longer wets the rock surface.

The application of the concept of wettability to reservoir rocks can be seen in the following description. The wettability of rock is influenced by the chemical composition of its constituents. So the concept of chemical element bonds greatly affects the wettability of a rock surface against water or oil. Wettability classification based on rock lithology is shown in the following table (**Table 1**).

For petroleum exploration and exploitation purposes, water-wet formations are easier to perform on oil recovery. This is because the water-wet formation has a high oil saturation, which means that water wets and fills the smallest pores in the rock grains and the oil is above the water so that it is easier for oil to move and spill because there is no adhesion force on the rock surface. This situation allows the increase in oil recovery in formations with sandstone reservoirs by injection of water into the formation so that it can increase the volume of water and encourage oil to come out and be produced.

Lithology	Wettability
Sandstone	Neutral-wet—water-wet
Clay	Neutral-wet
Carbonate	Neutral-wet—oil-wet

Table 1.
Wettability classification based on its lithology [24].

Wettability is one of the important parameters that affect the distribution of fluid in the reservoir and has a great influence on the spontaneous imbibition process. Wettability is defined as the ability of a fluid phase to wet certain solid surfaces while also facing another immiscible fluid [25]. The fluid is called as wetting phase and the other one is called the non-wetting phase. Wettability is a vital concept in oil recovery process because of its significant impact toward distribution ability, location determination and level of oil and water movements in the reservoir. In a water-wet system, water is expected to penetrate and stay inside narrow pores and drive oil out of the pore in the form of droplets. The opposite condition occurs in the oil-wet reservoir. This wettability can be determined based on the measurement results of the angle of contact; if it is from 0° to 75°, it is defined as water-wet, while if the angle is at 105–180°, it is defined as oil-wet, which makes angle range of 75–105° as neutral-wet system [26].

Capillary pressure is a pressure difference that occurs on interface of two unmixed fluids, where one fluid wets the surface of rock more than the other.

IFT also influences capillary pressure and will influence fluid distribution and flow. The ratio between viscosity and capillary forces is represented by capillary number (N_c) in the following formula:

$$N_c = (\upsilon\mu)/(\sigma) = (\text{viscous forces})/(\text{capillary forces}) \qquad (2)$$

where:
υ = average speed of fluid in pores (ft/D)
μ = viscosity of driving fluid (cP)
σ = interfacial tension system between water and oil (dyne/cm)

Surfactants can lower the IFT of oil-water, which results in the increase of N_c. IFT also acts as a guidance for miscibility; high IFT indicates immiscible fluid and low IFT indicates miscible fluid. The clear difference of low and ultralow IFT on surfactant flooding is detected on its emulsification approach. However, both w/o or o/w emulsions are formed on pored media, specifically on the angle of contact, in the form of droplets [27]. The contact angle observed during unmixed movement is not mixed in pored media, is arranged by equilibrium of capillary and viscosity forces, and escalates as the number of capillary increases. On surfactant injection with low IFT level in water-wet reservoir, the capillary number is inadequate to increase contact angle at a significant level, which creates a state where oil remains on non-wetting phase and leads to o/w emulsion formation (**Figure 1a**). On surfactant injection with ultralow IFT in water-wet reservoir, the high number of capillaries increases the angle of contact by more than 90° (**Figure 1c**) and creates w/o emulsion instead of o/w emulsion during movement (**Figure 1d**). The formation of a thick w/o emulsion on surfactant inundation with ultralow IFT leads to increased pressures and swiping efficiency, which eventually create an effective and improved crude oil production.

Surfactant injection is used to significantly decrease capillary forces in order to be able to mobilize the remaining oil after water injection. As the surfactant is injected into the reservoir, the tail of the surfactant will interact with oil residual; meanwhile, the head will interact with brine, which leads to lower IFT level [28]. Low IFT in the reservoir will eventually lead to reduced flow resistance and increased oil mobilization. Selection of the right surfactant is an important factor that causes reduced IFT and helps in the recovery of 10–20% of the original oil reserve.

The formation of microemulsions in surfactant injection in chemical flooding is very important to create effective injection. Microemulsion can be produced both through surfactant-oil mixing and surfactant injection to create in situ microemulsion.

A homogeneous mixture of oil and water while surfactant is present will increase oil movement in the reservoir. The effect will be lower IFT between oil and water, which leads to lower capillary pressure and mobilizes oil residual [29]. Under normal conditions, the reservoir is water-wet, so water tends to stick to the rock surface, while oil lies between the water phases. This is a normal condition where oil can flow by itself to be produced to the surface. In the case of tertiary recovery conditions, where oil is trapped in rock pores and cannot move on its own, in general the reservoir is oil-wet. In an oil-wet system, oil occupies a narrow pore and is present as a film on the pore wall, while water is present as water droplets in the middle of the pore. In this condition, the so-called oil droplets stick to the pore walls. To release the oil attached to the pore wall, a surfactant is needed, which functions to reduce the interfacial tension between the oil grains on the pore wall. In an oil-wet reservoir, oil residual stays inside rock pores, which makes it difficult to be mobilized except with high level of energy release. Surfactants act as an altering agent to change the wettability into water-wet [30]. The nature of SLS surfactant as an o/w emulsification allows it to convert oil-wet into water-wet.

Wettability is an important factor in EOR. After the wettability change occurs, the detachment energy is significantly reduced, making it easier to mobilize the remaining oil. Wettability is an important factor that determines the pattern of oil-water transfer in porous media. "Wettability" is a term to describe the relative adhesion of two fluids to a solid surface. In porous media filled with two or more immiscible fluids, wettability is a measurement of which fluid can wet (spread or adhere to) the surface. In a water-wet system (wet water), rocks filled with oil and water, water occupies the smallest pores and wets most of the surface in the larger pores. In areas that have high oil saturation, the oil is held above the wet water and spread on the surface. If the rock surface tends to be water-wet and the rock is saturated with oil, water fills the smallest pores, displacing oil when water enters the system. If the rock surface tends to be oil-wet, it is saturated with water; oil enters and wets the smallest pores in place of water. A rock saturated with oil means water-wet (wet water), and vice versa if the rock is saturated with water means oil-wet. By lowering the oil/water interfacial tension and changing the wettability from oil-wet to a water-wet condition, the recovery rate of the remaining oil can be increased successfully. The combined effect of wettability-IFT shows a better potential for EOR, compared to the reduction in interfacial tension alone under oil-wet conditions [31].

3. Results and Discussion

In this research, we utilized SLS surfactant synthesized from bagasse. SLS surfactant is anionic, which makes it popular in surfactant injection processes. The function is to drive middle-phase emulsion and lower IFT level of oil so that it can be mobilized and produced [28, 32]. Lignosulfonate is a derivative of lignin that contains hydrophilic groups, namely sulfonates group, hydroxyl alcohol, phenyl hydroxyl and hydrophobic groups [33]. That is the reason why this type of surfactant is categorized as anionic.

The phase behavior test was carried out to see changes in behavior from the condition of the crude oil sample, before and after being mixed with surfactant. The phase test was carried out in 2 mL of SLS surfactant mixed with 2 mL of crude oil sample in a 5 mL test tube. After entering the two types of fluid, the tubes are shaken slowly. The phase behavior test was carried out for 21 days at a constant temperature in the oven. In **Figure 8A**, there is still a clear boundary between the surfactant at the bottom and crude oil at the top. After being shaken slowly, in **Figure 8B** it can be seen that there is a mixture between the surfactant part and the

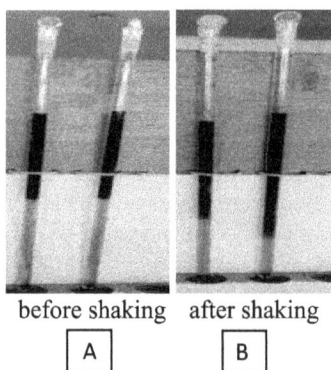

before shaking   after shaking

A   B

Figure 8.
Phase behavior test before and after shake.

crude oil part. This is referred to as a microemulsion or middle-phase emulation, because of the mixing that occurs at the boundary between the surfactant and crude oil. **Figure 8B** is the Winsor III phenomenon, as expressed in the microemulsion formation mechanism [14, 16]. The following figures picture phase behavior test before and after shaking.

The spontaneous emulsification shows that oil grains can be mixed with surfactant grains. Emulsion between sugarcane bagasse SLS surfactant and light oil shows the existence of a system that includes two liquid phases dispersed as small grains (**Figure 9**). The emulsion that forms in the middle of the pipette has shown the occurrence of Winsor III emulsion.

The phase behavior test results of a number of samples show that sugarcane bagasse SLS surfactant is able to form microemulsion with crude oil, as seen in the following figure.

Figure 10a shows phase behavior test on light crude oil sample; **Figure 10b** shows phase behavior test result on intermediate crude oil; and **Figure 10c** shows phase behavior test result on heavy crude oil samples. Sugarcane bagasse SLS surfactant will only form emulsion with light crude oil. The emulsion is not formed at all on intermediate and heavy crude oil samples. The phase behavior test result on light crude oil is continued with a number of different formation water salinity compositions (**Table 2**).

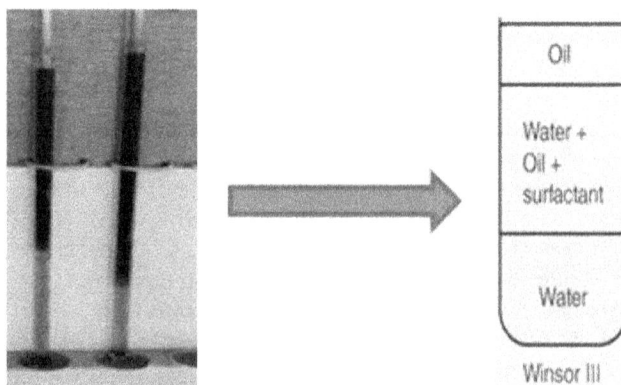

Figure 9.
Microemulsion formation (middle-phase emulsion).

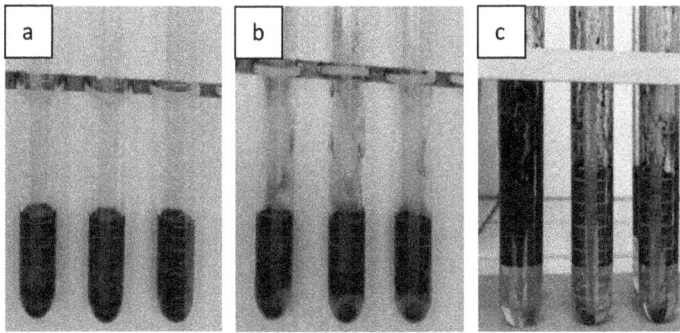

Figure 10.
Phase behavior test on a number of crude oil samples.

No.	Salinity (ppm)	Stability emulsion (%)	Type emulsion
1.	5000	10.00	Middle phase
2.	10,000	8.75	Middle phase
3.	20,000	5.00	Upper phase
4.	40,000	6.25	Middle phase
5.	80,000	1.25	Upper phase

Table 2.
Phase behavior test results on a number of salinities.

In some compositions with certain salinity, emulsions are formed in the middle and upper phases. Middle-phase emulsion or microemulsion occurs in the composition of sugarcane bagasse SLS surfactant with salinity of 5000 ppm, 10,000 ppm and 40,000 ppm, while the emulsion of the upper phase occurs at the composition of salinity of 20,000 ppm and 80,000 ppm. This is in accordance with the theory that states that the behavior of the microemulsion phase is very complex and depends on a number of parameters, including the type and concentration of surfactants, co-solvents, hydrocarbons, salinity of brine, temperature and much lower pressure levels [7]. In this case, formation water salinity influences microemulsion formation.

The formation of the mid-phase emulsion by the SLS surfactant is actually related to the components contained in the SLS surfactant. The results of the synthesis of SLS surfactants from bagasse have been analyzed using the Fourier transform infra-red (FTIR) test and the nuclear magnetic resonance (NMR) test. FTIR spectrophotometric measurements were carried out on the lignosulphonate surfactant product that had been produced from this lignin sulfonation process to determine the functional groups that corresponded to the expected lignosulphonate structure. The results of the FTIR test showed that the functional groups in the lignosulfonate structure of SLS bagasse consisted of —C=C-alkene groups, S=O sulfonates, C=O carboxylate groups and S—OR ester groups. This component complies with the standard lignosulfonate used as a comparison.

Furthermore, by using NMR, monomer test was carried out to determine the molecular formula of the sugarcane bagasse SLS surfactant. Based on the structure of the lignosulphonate monomer obtained from the NMR spectrum analysis, we can determine the lignosulphonate monomer by noticing the presence of specific atoms, namely C, O, H and S inside its structure. From the results of the NMR spectrum analysis, the number of atoms C = 11, O = 8, H = 16 and S = 1, so the empirical

Classification	Groups	Amount
Lipophilic group	=CH—	3
	—CH$_2$—	3
	—CH$_3$	2
Hydrophilic group	—SO$_3$Na	1
	—OH	3

Table 3.
Sugarcane bagasse SLS surfactant function group classification.

formula for lignosulphonate monomer is $(C_{11}H_{16}O_8S)_n$. Functional groups in their structure can be grouped as hydrophilic groups or lipophilic groups. The grouping can be seen as in the following table (**Table 3**).

By knowing the lipophilic group and the hydrophilic group, the value of the Hydrophilic-Lipophilic Balance (HLB) can be calculated. The HLB value can be used to estimate the properties of the SLS surfactant. The calculation of the HLB value refers to the Myers theory [34] with the following formula:

$$HLB = 20 \times (M_h)/(M_l + M_h) \tag{3}$$

where
M_h = weight of hydrophilic molecule group
M_l = weight of hydrophobic or lipophilic molecule groups
Based on the data from **Table 4**, we can calculate the weight of hydrophilic and lipophilic molecule groups.

$$M_h = (SO_3Na) + (OH) \times 3 = (32 + 48 + 23 + 51 = 154) \tag{4}$$

$$M_l = (CH) \times 3 + (CH_2) \times 3 + (CH_3) \times 2 = 111 \tag{5}$$

Based on that, the HLB value is at:

$$HLB = 20 \times 154/(111 + 154)) = 11.62 \tag{6}$$

HLB value of 11.62 is compared to Myers and Akzo Nobel tables as follows (**Table 4**).

The HLB value of this bagasse SLS bagasse surfactant, after being matched with the Myers and Azko tables, turned out to be in the o/w emulsification range. Thus, it can be said that the SLS sugarcane bagasse surfactant is perfect for its use as an o/w emulsion. In the injection process, surfactant is dissolved in formation water

Application	Range HLB	
	Myers	**Azko Nobel**
Defoaming of aqueous systems	—	1–3
W/O emulsification	2–6	3–6
Wetting and spreading	7–9	7–9
O/W emulsification	8–18	8–28
Detergency	3–15	12–15
Solubilization	15–18	11–18

Table 4.
Surfactant function category [34, 35].

and then injected into the reservoir in order to mobilize trapped oil on rock pores. In addition to its function as a driving force for oil in rock pores, this SLS surfactant will change the properties of the oil by reducing interfacial tension. The results of this study have proven that there is a decrease in IFT in the sugarcane bagasse SLS surfactant system to light crude oil as a continuation of the results of the phase behavior and microemulsion stability tests (**Table 5**).

From the results of this IFT test, the lowest value of the IFT measurement was formed in the composition of the SLS sugarcane bagasse surfactant with a salinity of 10,000 ppm, namely 2.73 m/Nm. This sample of crude oil X comes from an oil field that has 11,000 ppm formation water. The initial IFT measurement of formation water without SLS surfactant was at 12.43 mN/m. The following **table 5** describes the changes in IFT measurement results without surfactant and with SLS bagasse surfactant. This measurement shows that the surfactant function can reduce IFT from 12.43 to 2.73 mN/m. This condition is in accordance with the theory that states that when surfactant is injected into the reservoir, the hydrophobic surfactant tail will interact with the residual oil and the hydrophilic head will interact with the saltwater, as shown in **Figure 4**. It is this hydrophobic and hydrophilic interaction with water and oil that causes a strong decrease in IFT (**Figure 11**) [28].

The picture shows a mixture of oil and water, forming a new arrangement, as occurs between a mixture of sugarcane bagasse SLS surfactants and crude oil X. This new molecular arrangement involves interactions between the hydrophobic components of the surfactant and oil on one side of the interface, and water on the other. This mechanism occurs in the mixing of sugarcane bagasse SLS surfactant with the light crude oil sample used. The hydrophilic group of sugarcane bagasse SLS surfactant consisting of SO3Na + and OH- interacts with saltwater and the

No.	Solution composition	IFT (mN/m)
1.	Light oil with formation water (without sugarcane bagasse SLS surfactant)	12.43
2.	Light oil with sugarcane bagasse SLS surfactant of 5000 ppm salinity	6.81
3.	Light oil with sugarcane bagasse SLS surfactant of 10,000 ppm salinity	2.73
4.	Light oil with sugarcane bagasse SLS surfactant of 20,000 ppm salinity	4.13
5.	Light oil with sugarcane bagasse SLS surfactant of 40,000 ppm salinity	4.11
6.	Light oil with sugarcane bagasse SLS surfactant of 80,000 ppm salinity	3.61

Table 5.
IFT test results of light oil X—sugarcane bagasse surfactant SLS 1.5% on salinity variations.

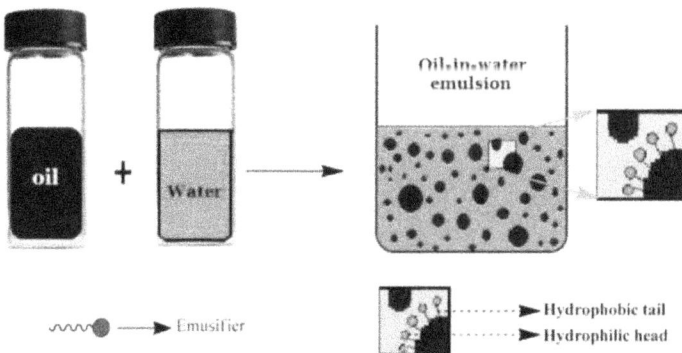

Figure 11.
O/W emulsion [36].

hydrophobic group consisting of $=CH-$, $-CH_2-$ and $-CH_3$ interacts with oil [36, 37]. With the result, we can say that oil-water interface is highly stronger than water-oil interaction prior to surfactant addition, which decreases IFT level [38, 39].

With the decrease in interfacial tension, the capillary pressure in the area of narrowing of the pores will decrease. Because the capillary pressure is reduced, the resistance to oil flow will also decrease so that the oil will move more easily to be produced [40, 41]. As the interfacial tension decreases, the capillary pressure decreases so the grain-fluid contact angle and the rock wall will also change. This has an impact on the wettability. Changes in wettability play a very important role in the mobilization of residual oil. In porous media filled with two or more immiscible fluids, wettability is a measurement of which fluid can wet (spread or adhere to) the surface [42]. In a water-wet system where the rocks are filled with oil and water, water will occupy the smallest pores and wet most of the surface in the larger pores. In areas that have high oil saturation, the oil will be held above the wet water and spread on the surface. If the rock surface tends to be water-wet and the rock is saturated with oil, water will fill the smallest pores, displacing oil when water enters the system [43]. If the rock surface tends to be oil-wet, it will be saturated with water, oil will enter and wet the smallest pores in place of water. A rock saturated with oil means water-wet (wet water) and vice versa if the rock is saturated with water means oil-wet. The wettability of a system can be classified in the very water-wet or oil-wet range depending on the water-oil interaction with the rock surface. The function of surfactants in the microemulsion formation mechanism will ultimately change the wettability of the reservoir system from water-wet to oil-wet.

4. Conclusions

Based on studies conducted on bagasse along with its utilization as a novel and stronger product, we can say that lignin content plays an important role that enables it to be produced as SLS surfactant that can be utilized on injection-based EOR process. This fact is also supported by laboratory test results, which evidently revealed that SLS surfactant contains four important compounds; alkenes, carboxylic acids, sulfonic acids and esters. As an addition, the liquid possesses HLB (-Hydrophilic-Lipophilic Balance) value of 11.62, which indicates its ability to dissolve in water and to form surfactant solution. The idea of SLS surfactant bagasse utilization is to create emulsion that would lower IFT level and mobilize oil better. That is why bagasse is highly potential to be utilized as main material of SLS surfactant production.

Author details

Rini Setiati*, Muhammad Taufiq Fathaddin and Aqlyna Fatahanissa
Petroleum Engineering, Faculty of Earth Technology and Energy, Trisakti
University, Jakarta, Indonesia

*Address all correspondence to: rinisetiati@trisakti.ac.id

IntechOpen

References

[1] Temizel C, Tuna T, Melih M, Oskay L, Saputelli A. Enhanced oil recovery and geothermal formulas and calculations. Formulas and Calculation For Petroleum, Engineering. 2019;**10**: 415-442. Available from: https://www.sciencedirect.com/science/article/pii/B978012816508900010X. DOI: 10.1016/B978-0-12-816508-9.00010-X

[2] Satter A, Iqbal GM. Enhanced oil recovery processes: thermal, chemical, and miscible floods. The Fundamentals, Simulation, and Management of Conventional and Unconventional Recoveries, Reservoir Engineering. 2016;**17**:313-337. Available from: https://www.sciencedirect.com/science/article/pii/B9780128002193000176. DOI: 10.1016/B978-0-12-800219-3.00017-6

[3] Gbadamosi AO, Junin R, Manan MA, Agi A, Yusuff AS. An overview of chemical enhanced oil recovery: Recent advances and prospects. International Nano Letters. 2019;**9**:171-202. Available from: https://link.springer.com/article/10.1007/s40089-019-0272-8

[4] Acosta E. Modeling and Formulation of Microemulsions: The Net-Average Curvatre Model and The Combine Linker Effect. Norman: University of Oklahoma; 2004

[5] Nedra Karunaratne D, Pamunuwa G, Ranatunga U. Introductory chapter: Microemulsions. In: Properties and Uses of Microemulsions. IntechOpen; 2017. DOI: 10.5772/intechopen.68823. Available from: https://www.intechopen.com/chapters/55232

[6] Kanwar R, Rathee J, Patil MT, Mehta SK. Microemulsions as nanotemplates: A soft and versatile approach. In: Microemulsion—A Chemical Nanoreactor. IntechOpen; 2018. DOI: 10.5772/intechopen.80758. Available from: https://www.intechopen.com/chapters/63349

[7] Sheng JL. Chemical Flooding in Modern Chemical Enhanced Oil Recovery. Oxford: Elsevier; 2011. pp. 239-335

[8] Babakhani P, Azhdarpour A, Zare A. Simulation study of micellar/polymer flooding process in sandpack. In: Proceedings of the 2011 IAJC-ASEE International Conference. 2011. Available from: https://ijme.us/cd_11/PDF/Paper%20239%20ENG%20108.pdf

[9] Labrid JC. Oil displacement mechanism by Winsor's type i micellar systems. In: SPE Annual Fall Technical Conference and Exhibition. 1979

[10] Holmberg K, Shah DO, Schwuger MJ. Handbook of Applied Surface and Colloid Chemistry. Vol. 1. United Kingdom: John Wiley & Sons; 2002

[11] Lawrence MJ, Rees GD. Microemulsion-based media as novel drug delivery systems. Advanced Drug Delivery Reviews. 2000;**45**:89-121

[12] McClements DJ. Nanoemulsions versus microemulsions: terminology, differences, and similarities. Soft Matter. 2012;**8**:1719-1729. DOI: 10.1039/c2sm06903b

[13] Winsor PA. Hydrotropy, solubilisation and related emulsification processes. Journal of Transaction Faraday Society. 1948;**44**:376-398

[14] Gautam N, Kesavan K. Development of microemulsions for ocular delivery. Therapeutic Delivery. 2017;**8**(5):313-330. DOI: 10.4155/tde-2016-0076

[15] Gosh P. Emulsion, Micromulsion and Foams. India: Chemical Engineering, IIT Guwahati; 2014

[16] Xueyu D, Lucia LA, Ghiladi RA. A novel approach for rapid preparation

of monophasic microemulsions that facilitates penetration of woody biomass. ACS Sustainable Chemistry and Engineering. 2016;**4**(3):1665-1672. DOI: 10.1021/acssuschemeng.5b01601

[17] Malik MA, Wani MY, Hashim MA. Microemulsion method: A novel route to synthesize organic and inorganic nanomaterials. Arabian Journal of Chemistry. 2012;**5**(4):397-417. DOI: 10.1016/j.arabjc.2010.09.027

[18] Craig JF. The Reservoir Engineering Aspects of Waterflooding. Tulsa: Society of Petroleum Engineers; 1993

[19] Munir R, Syed HK, Asghar S, Khan IU, Rasul A, Irfan M, et al. Microemulsion: promising and novel system for drug delivery. Journal of Toxicological & Pharmaceutical Sciences. 2017;**1**(2):128-134. Available from: https://www.researchgate.net/publication/325216233

[20] Gasco M, Gallarate M, Trotta M, Bauchiero L, Gremmo E, Chiappero O. Microemulsions as topical delivery vehicles: ocular administration of timolol. Journal of Pharmaceutical and Biomedical Analysis. 1989;7:433-439

[21] Kreilgaard M. Influence of microemulsions on cutaneous drug delivery. Advanced Drug Delivery Reviews. 2002;**54**:S77-S98

[22] Gadhave AD, Waghmare JT. A short review on microemulsion and its application in extraction of vegetable oil. International Journal of Research in Engineering and Technology. 2014;**3**:147-158

[23] Yuan Y, Randall Lee T. Contact angle and wetting properties. Surface Science. 2013;**51**:3-34

[24] Donalson EC, Thomas RD, Lorent PB. Wettability determination and its effect on recovery efficiency, SPE-2338-

PA. Society of Petroleum Engineers Journal. 1969;**9**(01):13-20

[25] Falode O, Manue E. Wettability effects on capillary pressure, relative permeability, and irredcucible saturation using porous plate. Journal of Petroleum Engineering. 2014;**2014**: 465418. DOI: 10.1155/2014/465418

[26] ElMofty O. Surfactant Enhanced Oil Recovery by Wettability Alteration in Sandstonereservoirs. Missouri: Student Research & Creative Works at Scholars' Mine, Missouri University of Science and Technology; 2012

[27] Guba S, Horváth B, Szalai I. Examination of contact angles of magnetic fluid droplets on different surfaces in uniform magnetic field. Journal of Magnetism Magnetic Materials. 2020;**498**:166181. DOI: 10.1016/j.jmmm.2019.166181

[28] Sheng JJ. Status of surfactant EOR technology. Petroleum. 2015;**1**(2): 97-105. DOI: 10.1016/j.petlm.2015.07.003

[29] Alli YF, Tobing EML, Usman U. Microemulsion flooding mechanism for optimum oil recovery on chemical injection. Scientific Contribution Oil & Gas Journal. 2017, 2017;**40**(2):85-90. Available from: http://journal.lemigas.esdm.go.id/index.php/SCOG/article/view/43/pdf

[30] Yang W, Brownlow JW, Walker DL, Lu J. Effect of surfactant-assisted wettability alteration on immiscible displacement: A microfluidic study. Water Resources Research. 2021;**57**(8). DOI: 10.1029/2020WR029522

[31] Zhao J, Wen D. Pore-scale simulation of wettability and interfacial tension effects on flooding process for enhanced oil recovery. RSC Advances. 2017;7:41391-41398. DOI: 10.1039/C7RA07325A

[32] Bera A, Mandal A. Microemulsions: A novel approach to enhanced oil recovery: A review. Journal of Petroleum Exploration and Production Technology. 2015;**5**:255-268. Available from: https://link.springer.com/article/10.1007/s13202-014-0139-5

[33] Giang NT, Kien TT, Hoa NT, Van Thiem P. A new synthesis process of lignosulfonate using lignin recovered from black liquor of pulp and paper mills. Journal of Science and Technology. 2016;**54**(4B):1-10. DOI: 10.15625/2525-2518/54/4B/12017

[34] Myers D. Surfactant Science and Technology. New Yersey: Wiley Interscience; 2006. Available from: http://press.crosa.com.tw/wp-content/uploads/2016/11/Surfactant-Science-and-Technology-Third-Edition.pdf

[35] Nobel A. HLB & Emulsification. Chicago, USA: Surface Chemistry LLC; 2011

[36] Elanchezhiyan SSD, Prabhu SM, Meenakshi S. Effective adsorption of oil droplets from oil-in-water emulsion using metal ions encapsulated biopolymers: Role of metal ions and their mechanism in oil removal. International Journal of Biological Macromolecules. 2018;**112**:294-305. DOI: 10.1016/j.ijbiomac.2018.01.118

[37] Setiati R, Siregar S, Marhaendrajana T, Wahyuningrum D. Influence of middle phase emulsion and surfactant concentration to oil recovery using SLS surfactant synthesized from bagasse. IOP Conf. Series: Earth and Environmental Science. 2018;**212**: 012076. DOI: 10.1088/1755-1315/212/1/012076

[38] Azodi M, Nazar ARS. Experimental design approach to investigate the effects of operating factors on the surface tension, viscosity, and stability of heavy crude oil-in-water emulsions. Journal of Dispersion Science and Technology. 2013;**34**:273-282. DOI: 10.1080/01932691.2011.646611

[39] Xu J, Zhang Y, Chen H, Wang P, Xie Z, Yao Y, et al. Effect of surfactant headgroups on the oil/water interface: An interfacial tension measurement and simulation study. Journal of Molecular Structure. 2013;**1052**:50-56. DOI: 10.1016/j.molstruc.2013.07.049

[40] Hu G, Ma D, Wang H, Wang F, Yuanyuan G, Yu Z, et al. Proper use of capillary number in chemical flooding. Journal of Chemistry. 2017;**2017**: 4307368, 11 pages. DOI: 10.1155/2017/4307368

[41] Seng KK, Loong WV. Introductory chapter: From microemulsions to nanoemulsions. In: Nanoemulsions—Properties, Fabrications and Applications. IntechOpen; 2019. DOI: 10.5772/intechopen.87104

[42] Dehkordi et al. A field scale simulation study of surfactant and polymer flooding in sandstone heterogeneous reservoir. Journal of Petroleum & Environmental Biotechnology. 2018;**9**:1. DOI: 10.4172/2157-7463.1000366

[43] Behera MR, Varade SR, Ghosh P, Paul P, Negi AS. Foaming in micellar solutions: Effects of surfactant, salt, and oil concentrations. Industrial & Engineering Chemistry Research. 2014; **53**(48):18497-18507. DOI: 10.1021/ie503591v

Interfacial Behavior of Saponin Based Surfactant for Potential Application in Cleaning

Gajendra Rajput, Niki Pandya, Darshan Soni,
Harshal Vala and Jainik Modi

Abstract

Amphiphilic molecules having a tendency to decrease the surface tension of the aqueous medium and those are widely used in the industrial and domestic sector. Nowadays they are in high demand to replace synthetic surfactants by naturally based molecules to reduce the environmental problem. Approx. more than 60% materials which are of surfactants-based enter into the marine water which dangerous for aquatic lives. We propose novel material which is a natural based surfactant which is biodegradable and eco-friendly alternatives. Here we are focused on tea saponin and investigated properties like surface tension, foaming, skin mildness, cleaning ability. This is maybe the first reporting a surfactant activity of tea based surfactant. Natural originated surfactants display well emulsion making capacity at the large amount as compared to synthetic. Tea is acidic in nature and it reduces surface tension to 31.4 mN/m, and greater foam, ultra-mildness, with excellent cleaning ability. The consequences suggest that tea have outstanding surface-activity which can be used as a green replacement for synthetic surfactants.

Keywords: green surfactant, saponin, foaming, emulsification, mildness

1. Introduction

Amphiphilic molecules contain hydrophilic and hydrophobic units that improve interfacial properties. This characteristic of surfactant makes it suitable for the fields of detergent, wetting, emulsification, oil recovery, froth flotation and other fields [1–3]. A large number of non-natural surfactants are used in industrial and domestic work which are spread over underwater, soil, sediment, etc. Studies have shown that more than 60% of surfactant derivatives are released into the aquatic environment. Global surfactant production was approximately 12.5 million tons in 2006 [4], while production in Western Europe that is more than 3 million tons in 2007 [5]. After that year of 2010, the use of non-ionic surfactants (polyethoxylated nonylphenol) in the United States was approximately 172,000 tons [6]. These synthetic surfactants can disturb the environment and cause health risks such as respiratory tract, dermatitis, eye irritation, etc. [7, 8].

Ostroumov expressed how non-biodegradable cleaning product can reduce the cleaning capacity of oysters and mussels. Reduce the Water purifying capacity of bacteria like *Crassostrea gigas* and *Mytilus galloprovincialis* which cause an important impact on the ecosystem. Besides, aquatic plants, microorganisms, humans, also affected by the non-biodegradable surfactants [9].

Due to technological advancement methods for the mass production of surfactants are emerging, which have led to serious environmental problems [10]. Today's demand, surfactants should be biodegradable and have minimal toxicity to be surface active. Therefore, researchers are looking for environment-friendly surfactants, it may be a natural surfactant. Natural materials are gained from natural sources, such as plant, bacterial or fungal, animal fats. Various methods are needed, for extract and using them. The fatty acid esters or amides of these sugars can be used as substitutes for synthetic surfactants, as described by Salati et al. Humic acid extraction using biomass as a surfactant has been reported [11].

Saponins are well known plant-based surfactants. Natural surfactants have biodegradability, biocompatibility, and low toxicity, so they are less harmful to the environment [12]. These products can be manufactured with high production with cost effect which could be used in environmental control actions such as processing industrial emulsions, controlling oil pollution, detoxifying industrial wastewater, and bioremediation of contaminated soil [13].

Here we report that tea saponin is considered as a natural surfactant. A recent movement in the industry to rejecting used of synthetic products which inspiration for people to look for natural-based new materials. By these inspirations, we examined tea to estimate its physiochemical property such as critical micelle concentration (CMC), Foaming, viscosity behavior, emulsification, mildness pH, and conductivity. For comparison, we also examined the marketed product (ionic surfactants) just because they are commercially availability. We can reduce environmental issues just by replacing them with natural surfactant.

2. Experimental procedure

2.1 Materials

Tea saponin powder was purchased from King diamond (65% approx.). Surf excel (Hindustan Unilever Limited), as a marketed surfactant, Soya bean oil (refined, batch number (AF) SB06C04), Coconut oil (Lot No. KB003), Paraffin wax (candles from the market) were used. Hexane was obtained from Loba Chemie. Milli-Q water of surface tension 71.6 ± 0.1 mN/m was used. Bleached cotton cloth Sort No. 22425003 from Akash Textiles, Ahmedabad.

2.2 Surface tension measurements

Surface tension of the individual surfactant was determined by a force tensiometer (*Type: K20, KRÜSS, Germany*) using a platinum plate method. The experiment was directed by using 25 mL of surfactant solution in a sample container. In this method, the liquid to be inspected is placed in a container, and its position can be changed by using a screw to move up and down. There is a platinum plate on the top of the sample. During this process, to ensure the plate is accurately immersed in the liquid alignment of platinum plate adjusted at 3 mm distance from the liquid. In the tensiometer, a pre-programmed software archives the surface tension of liquid under examination. For accuracy before apiece measurement instrument was calibrated using Milli-Q water (71.5 ± 0.1 mN/m) at 27 ± 0.5°C. After the completion of each reading, the platinum plate was carefully washed using deionized distilled water, followed by heating to ensure that the water evaporated from the surface of the plate. All measurements were recurrent three times to accord repeatability of the data.

2.3 Foam ability and stability

The Ross Miles method was utilized to study the foaming behavior of surfactant. To study the foaming behavior 250 mL surfactant solution was prepared. Primary step to perform foam analysis was to wash the column with deionized water followed by rinsing of the column using surfactant solutions under examination. Afterwards, from 250 mL prepared surfactant solution gradually 50 mL of surfactant solution poured into the column and was stored at the bottom part of column. The remaining 200 mL surfactant solution was taken into the pipette and pipette containing surfactant solution positioned at top of the column. Then, faucet of pipette opened and at fixed flowrate surfactant solution run off into the column. As soon as the surfactant solution from pipette completely transfer to column, start the stop watch and note the initial foam height (termed as foamability) and after 5 min note foam height again (termed as foam stability).

2.4 pH and conductivity

pH and conductivity were measured by Laqua pH 1100 (Horiba) at room temperature. The pH and conductivity electrode was calibrated with the help of standard solutions previously measurements.

2.5 Viscosity

Vibration viscometer (SV-10 series, A & D Company, Limited) was employed for viscosity measurement of samples. The working principle of instrument to determine viscosity followed constant parameters like frequency 30 Hz and amplitude <1 mm, at these fixed parameters the required current to resonate sensor plates used to detect viscosity.

2.6 Emulsification

This measurement carried out by the Kothekar method [14]. Initially 10 mL of sample solution and 10 mL of refined oil was taken into a graduated cylinder followed by shaking of this mixed solution. Now for this emulsion time required to separate 5 mL of sample solution was determined and denoted as emulsion persistence.

2.7 Cleaning

To determine the cleaning efficiency the Sharma method [15] was considered, which involved four steps. Firstly 5 × 5 cm sized cotton cloth was soaked in water for 12 h, dried and weighed. In second step simulate dirt was prepared by maxing 0.5 g coconut oil and 0.5 g paraffin in 50 mL of hexane. Now, in third step the dry cotton cloth was two times dipped into the simulated oil, dried and weighed. Lastly the cloth was treated with surfactant by soaking it for 10 min followed by washing, drying and weighing. To determine cleaning efficiency different surfactant concentrations ranging from 0.01% to 0.1% were utilized at room temperature.

2.8 Protein solubilization

The protein (zein) solubilization potential of surfactant was quantified by the gravimetric analysis. The 1.0% by weight surfactant solutions were prepared in vials and approximately 2.0% by weight zein powder added to the solution. Now these mixtures of surfactant-zein protein were continuously stirred for 24 h. After 24 h continuous stirring the solutions were filtered using Whatmann filter paper to collect insoluble zein from solution. Lastly the collected insoluble solid zein was desiccated at 80°C for 24 h and weighed. Using the insoluble zein powder weight the solubilized zein% has been calculated.

2.9 Lipid solubilization

The lipid (steric acid) solubilization potential of surfactant was quantified by the gravimetric analysis. The 1.0% by weight surfactant solutions were prepared in vials and approximately 2.0% by weight stearic acid powder added to the solution. Now these mixtures of surfactant-stearic acid were continuously stirred for 24 h. The surfactant-stearic acid mixture was then filtered using Whatman filter paper. Finally, the collected solid was dried at 60° C for 24 h. The weight percentage of the solubility of the lipid in the surfactant was calculated from the weight of the insoluble stearic acid after drying for 24 h.

3. Microstructure

Microstructure of gas bubbles was measured through an Olympus STM7CB Digital microscope. The foam was generated by using the handshaking method.

4. Results and discussion

4.1 Surface tension

The surface tension creates an imbalance in the intermolecular force in an interface between liquid, vapor or liquid and solid. **Figure 1** shows that the reduction in surface tension due to surfactant concentration increment and after certain concentration surface tension value was constant as this concentration is called the critical concentration of micelles (CMC). The reduction in surface tension is due to the break of hydrogen bonds aqueous, the reason was increased adsorption of monomer

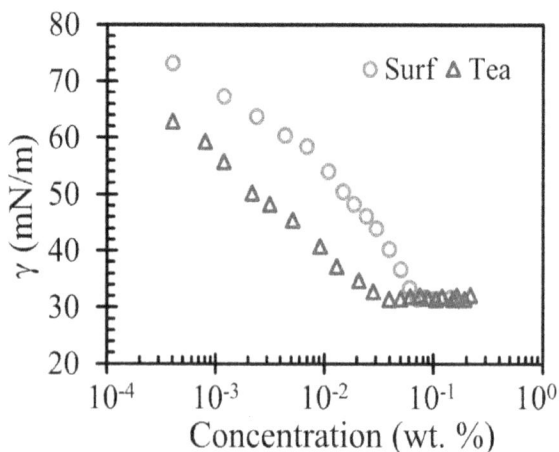

Figure 1.
Plot of equilibrium surface tension versus total surfactant concentration at a temperature of 27 ± 0.5°C.

Name of surfactant	CMC (wt%)	γCMC	πCMC
Surf excel (surf)	0.07	31.5	40.5
Tea saponin (tea)	0.04	31.4	40.6

Table 1.
Surface properties of surfactants.

at the air-water interface [1]. The surface tension reduction value between 32 and 37 mN/m consider as the material has good detergency and surface activity [16]. Therefore, tea has a decent washing ability.

The properties of surfactants, such as detergents, solubilizers, etc., depending on their structure. The reduction in surface tension caused by natural surfactants is almost close [17] to the surface tension obtained by measuring surface tension (**Table 1**). The ionic surfactants tend to have relatively greater CMC value as of non-ionic surfactant (as observed for surf), such behavior of ionic surfactants may be attribute to the repulsion between neighboring analogous head group charge. The tea saponin reveal greater capability to reduce surface tension as compared to surf may be due to absence of elementary electrostatic interaction in tea saponin and this also resembles faster micelle formation in water.

The effectiveness of surfactant (π_{CMC}) is given by π_{CMC} where γ_0 is the surface tension of pure water and γ_{CMC} is the surface tension of solution at CMC. Natural surfactant shows maximum effectiveness followed by synthetic.

$$\pi_{CMC} = \gamma_0 - \gamma_{CMC} \tag{1}$$

The dynamic surface tension of surfactants solutions was studied by the pendant drop method. **Figure 2** displays the dependencies of dynamic surface tension for micellar solutions of natural and synthetic as functions of the effective lifetime [18]. The concentrations of surfactant are 0.005 wt% (below CMC). From the result shows that the natural surfactant leads to faster the surface tension changes because the Natural surfactant having no kind of change repulsive force at a head group like surf.

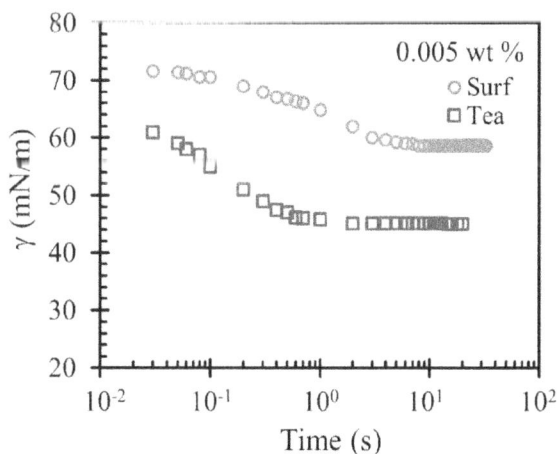

Figure 2.
Plot of surface tension as a function of time at a temperature of 27 ± 0.5°C.

4.2 Foam ability and stability

Foaming behavior and cleaning action is not much interrelated, but foam behavior is a significant condition in cleansing agent assessment by the consumers. Foam creation and durability are imperative in numerous applications [19–21]. Foam ability is the amount of foam creation due to the constant formation of new interfaces. Higher power of foaming requires faster adsorption, high surface elasticity. The foam results obtain by Ross Miles test are shown in **Figure 3** at a concentration range from 0.02 to 0.10 wt% for both surfactant sample. Foam generation by pouring method and the amount of foam formation was considered as Foam ability and after 5 min foam height measures that are termed as foam stability. The tea solution produces dense, high-quality foam, which may be due to the presence of high amounts of saponin. The existence of saponin group contributes to significantly reduce the dynamic surface tension as well as supports to produce the large surface for foam formation [13]. Foam formation upsurges with increased surfactant concentration as greater number of monomers could custom in film to enhanced foam stability.

We have also observed under the microscope to get an insight into the bubble size and foam structures as depicted in **Figure 4**. The foam engendered by surf solution speckled to contain more liquid portion with larger bubble size, while the foam generated using tea solution contained less liquid portion with smaller bubble size. These experimental results lead to better understanding of foam formed using natural surfactant and it indicates that such natural surfactants could deliberately consider as foaming agent for various applications. The foaming behavior studied by utilizing the Ross Miles method [22] and similar results were observed for foam generated by hand shake method as indicated in a picture (**Figure 4**) at 0.1 wt% concentration for both tea saponin and surf.

4.3 Viscosity

Micellization affects the viscosity of a solution, depending on the size and number of particles in the solution. **Figure 5a** shows that the viscosity gradually increases with increasing concentration. The absence of charge repulsion in between the head groups of surfactant monomers may lead to induce viscosity with

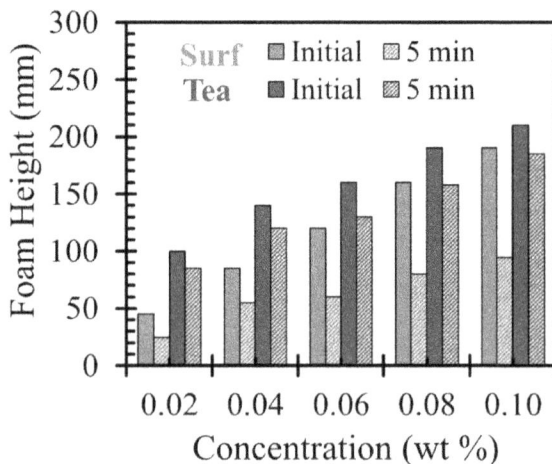

Figure 3.
Plot of foam height as a function of surfactant concentration in Milli-Q water at a temperature of 27 ± 0.5°C.

(a)

(b)

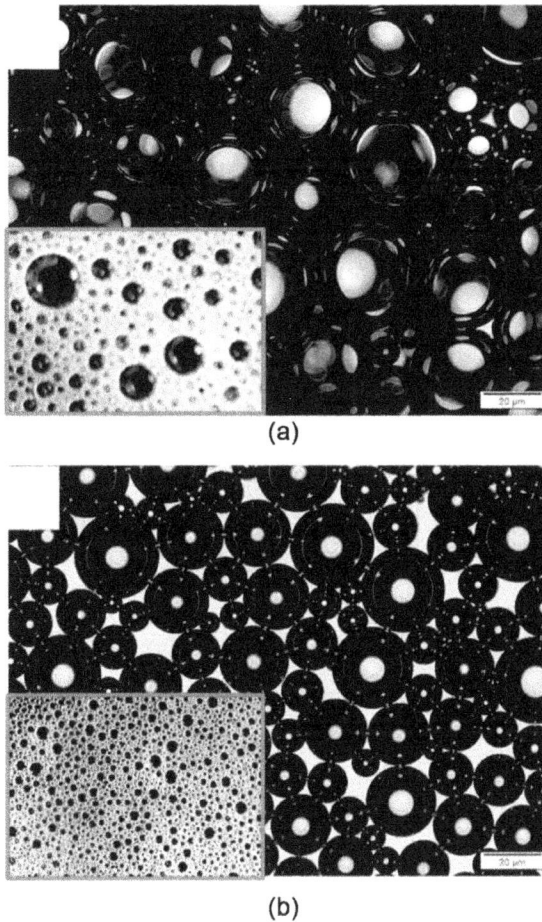

Figure 4.
Optical microscopy images of the foam generated from (a) surf (b) tea respectively at 27°C. Inset shows the photo of the foam for the corresponding surfactants.

increased concentration of tea, The hydrophilic part of surfactant monomers surrounded by water assists to a rise in viscous resistance. The viscosity progress may escalate rapidly above CMC due to micelle shape transition at higher concentration. The increase in viscosity far exceeds CMC, which is due to more interactions among micelles, the interactions among micelles are starting to get closer and reduce the critical packing parameters [23].

4.4 Conductivity

Detergent mostly made by ionic surfactants which are ionized in an aqueous condition so it will show conductance from low to high depends on concentration. The change in conductivity at lower concentrations almost constant further increases at higher concentrations for natural surfactant **Figure 5b**. At lower surfactant concentration, the headgroups of surfactant were encircled by water consequently results to lower conductivity. The conductivity upsurges with the increasing concentration of surfactant due to ionization of surfactant molecules [24]. The conductivity also associates with mildness, which demonstrates inversely

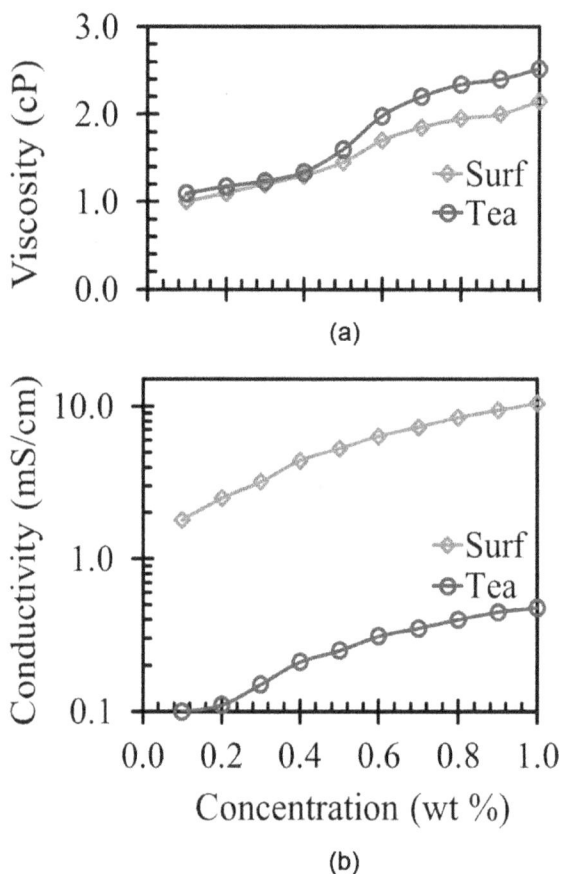

Figure 5.
Plot of (a) viscosity and (b) conductivity as a function of surfactant concentration at a temperature of 27 ± 0.5°C.

proportional correlation [25]. The experimental outcome indicates that the conductivity increase at higher concentration for tea ensues to be more as compared to surf.

4.5 pH measurements

In surfactant science research investigation of pH is an indispensable study. The pH of a surfactant solution relies on the overall charge of the headgroup, which consequently changes the repulsion between headgroups [7]. The pH values for both tea and surf solutions for range of concentration 0.02–0.10 wt% in Milli Q water were observed to be 6.7 ± 0.2. The pH values of the surfactant solution at various concentrations are shown in **Table 2**. Tea with an acidic pH, possibly due to the hydrolysis of nonionic glucuronic groups. This pH of the solution near the skin (~5.5) causes less damage to the hair and skin. The surf shows alkaline in nature. The pH of the tea solution decreases with concentration, while that of surf increases.

4.6 Emulsification

The emulsion is a fine dispersion of one liquid into another stabilized by emulsifier such as protein, surfactant, polymers etc. The surfactants can be solubilized non-polar substances into polar due to its amphiphilic nature, e.g. Surfactant

Concentration (wt%)	Surf	Tea
0.02	10.2	5.3
0.04	10.3	5.2
0.06	10.5	5.5
0.08	10.8	5.9
0.10	11.2	6.1

Table 2.
The values of pH at different concentrations of surfactants solutions.

monomers adsorb between the water-oil interface and reduce its interfacial tension. Due to this reduction in interfacial tension less energy is required to form new interfaces that are essential to the preparation of a stable emulsion. In these studies, we make simple oil-in-water (o/w) emulsion, which showed the emulsifying power of the surfactant's solution **Figure 6**. Emulsion stability increases with concentration increases, natural surfactant formed more stable (almost 30 min) emulsions at high concentration. The results of surface tension show that the stability of the emulsion decreases in the area of micelle formation, possibly because less surfactant is adsorbed at the oil-water interface. Tea has the best emulsion stability, followed by the surf. Stable emulsions occur when the adsorbed surfactant makes repulsive interactions among the drops and generates an energy barrier against breakage.

4.7 Cleaning

Cleansing activity means, the removal of unwanted substances such as soil, grease and dirt, is the main target of any detergent. The cleaning activity was calculated by the below equation

$$C = \left[(W_2 - W_3) / (W_2 - W_1) \right] \times 100\% \qquad (2)$$

where W_1 = initial weight of cloth, W_2 = weight of the cloth with simulated dirt and W_3 = weight after cleaning with surfactant solution and water. As shown

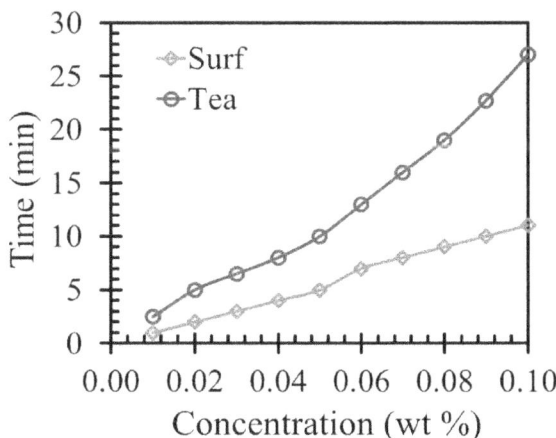

Figure 6.
Plot of emulsification as a function of surfactant concentration at a temperature of 27 ± 0.5°C.

in **Figure 7** The cleaning ability of the tea showed better at lower concentrations. The cleaning ability increases with concentration increases for natural surfactant. Although all surfactants have similar trends in cleaning performance, there is a significant difference in the amount of soil removed. The tea demonstrations decent cleaning efficacy at high concentration as compared to the surf, probably due to tea revealed greater efficiency to reduce surface tension.

4.8 Skin mildness of surfactants

In the skin structure first part is known as epidermis and the outer most layer of epidermis is stratum corneum (SC), this layer provides an important barrier function for skin. Surfactant can amend the function of SC by interacting with proteins and lipids of SC. These interactions lead to swelling and denaturation, however the comprehensive mechanism involved for such interaction has not been reported yet. But, predisposition of surfactant to interact with proteins could relate with its impact on human skin, generically termed as mildness to human skin [26].

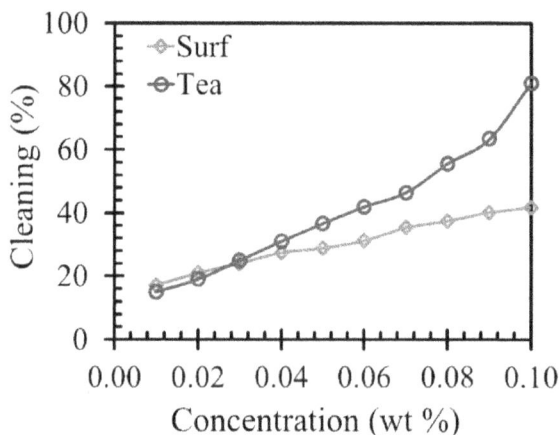

Figure 7.
Plot of cleaning ability as a function of surfactant concentration at a temperature of 27 ± 0.5°C.

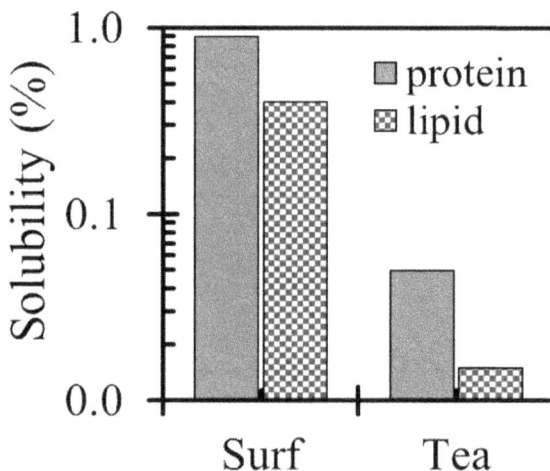

Figure 8.
Plot of protein and lipid dissolution by surfactants (1.0 wt%) at a temperature of 27 ± 0.5°C.

To comprehend impact of tea saponin and surf on skin, the solubilization potential of protein and lipid were determined using 1.0 wt% surfactant solution by dissolving model protein zein and model lipid stearic acid represented in **Figure 8**. The dissolution tendency of natural surfactants, zein and stearic acid is small compared to surf. This indicates that natural surfactants are milder than synthetic.

5. Conclusions

Herbal saponins, tea, were studied to find alternatives to synthetic surfactants that are commonly used and were make comparison with marketed available surfactant (Surf Excel). The outcomes demonstrated that saponins are naturally acidic and decomposable. Natural materials are considered to be biodegradable as plant extracts. The tea, which was probably examined by the first time, shows good effectiveness, in addition to high foaming capacity, decent cleaning capacity and ultra-soft. Although lots of works account on isolation, characterization, etc. Tea has good emulsifiers and can find some industrial uses. We also quantified the cleaning capacity of surfactant solutions. Therefore, our studies can offer a simple and inexpensive method to measure the general cleaning method for the evaluation of detergents. We conclude that tea comes with good surface-active properties. These studies can offer useful information for the food industry as well as the cosmetics industry reason was that these plant-based materials were biodegradable organic surfactant. We propose new biodegradable and renewable alternative from plant-based material which act as a surfactant.

Author details

Gajendra Rajput*, Niki Pandya, Darshan Soni, Harshal Vala and Jainik Modi
School of Engineering and Applied Science, Ahmedabad University,
Ahmedabad, India

*Address all correspondence to: gajendra.rajput@ahduni.edu.in

IntechOpen

References

[1] Rosen MJ, Kunjappu JT. Surfactants and Interfacial Phenomena. Hoboken, New Jersey, United States: John Wiley & Sons; 2012. DOI: 10.1002/978111822 8920

[2] Kronberg B, Holmberg K, Lindman B. Surface Chemistry of Surfactants and Polymers. Hoboken, New Jersey, United States: John Wiley & Sons; 2014. DOI: 10.1002/9781118695968

[3] Mittal KL. Solution Chemistry of Surfactants. Berlin/Heidelberg, Germany: Springer Science & Business Media; 2012

[4] Edser C. Latest market analysis. Focus on Surfactants. 2006;**5**:1-2. DOI: 10.1016/S1351-4210(06)71142-8

[5] Ivanković T, Hrenović J. Surfactants in the environment. Archives of Industrial Hygiene and Toxicology. 2010;**61**:95-110. DOI: 10.2478/10004-1254-61-2010-1943

[6] Schmitt C, Grassl B, Lespes G, Desbrières J, Pellerin V, Reynaud S, et al. Saponins: A renewable and biodegradable surfactant from its microwave-assisted extraction to the synthesis of monodisperse lattices. Biomacromolecules. 2014;**15**:856-862. DOI: 10.1021/bm401708m

[7] Tmáková L, Sekretár S, Schmidt Š. Plant-derived surfactants as an alternative to synthetic surfactants: Surface and antioxidant activities. Chemical Papers. 2016;**70**:188-196. DOI: 10.1515/chempap-2015-0200

[8] Pradhan A, Bhattacharyya A. Shampoos then and now: Synthetic versus natural. Journal of Surface Science and Technology. 2014;**30**:59-76. DOI: 10.18311/JSST/2014/1856

[9] Ostroumov S. Studying effects of some surfactants and detergents on filter-feeding bivalves. In: Aquatic Biodiversity. Berlin/Heidelberg, Germany: Springer; 2003. pp. 341-344. DOI: 10.1007/978-94-007-1084-9_24

[10] Chevalier Y. New surfactants: New chemical functions and molecular architectures. Current Opinion in Colloid & Interface Science. 2002;**7**: 3-11. DOI: 10.1016/S1359-0294(02) 00006-7

[11] Salati S, Papa G, Adani F. Perspective on the use of humic acids from biomass as natural surfactants for industrial applications. Biotechnology Advances. 2011;**29**:913-922. DOI: 10.1016/j.biotechadv.2011.07.012

[12] Mondal MH, Malik S, Garain A, Mandal S, Saha B. Extraction of natural surfactant saponin from soapnut (*Sapindus mukorossi*) and its utilization in the remediation of hexavalent chromium from contaminated water. Tenside Surfactants Detergents. 2017;**54**:519-529. DOI: 10.3139/113. 110523

[13] Hajimohammadi R, Hosseini M, Amani H, Darzi GN. Experimental design procedure for optimization of saponin extraction from *Glycyrrhiza glabra*: A biosurfactant for emulsification of heavy crude oil. Tenside Surfactants Detergents. 2017;**54**:308-314. DOI: 10.3139/113.110506

[14] Kothekar SC, Ware AM, Waghmare JT, Momin S. Comparative analysis of the properties of Tween-20, Tween-60, Tween-80, Arlacel-60, and Arlacel-80. Journal of Dispersion Science and Technology. 2007;**28**:477-484. DOI: 10.1080/01932690601108045

[15] Sharma P. Cosmetics: Formulation, Manufacturing and Quality Control. Delhi, India: Vandana Publications; 2008

[16] Mainkar A, Jolly C. Evaluation of commercial herbal shampoos.

International Journal of Cosmetic Science. 2000;**22**:385-391. DOI: 10.1046/j.1467-2494.2000.00047.x

[17] Negm NA, Mohamed AS. Surface and thermodynamic properties of diquaternary bola-form amphiphiles containing an aromatic spacer. Journal of Surfactants and Detergents. 2004;**7**:23-30. DOI: 10.1007/s11743-004-0284-z

[18] Miller R, Aksenenko E, Fainerman V. Dynamic interfacial tension of surfactant solutions. Advances in Colloid and Interface Science. 2017;**247**:115-129. DOI: 10.1016/j.cis.2016.12.007

[19] Varade D, Carriere D, Arriaga L, Fameau A, Rio L, Langevin E, et al. On the origin of the stability of foams made from catanionic surfactant mixtures. Soft Matter. 2011;**7**:6557-6570. DOI: 10.1039/C1SM05374D

[20] Bureiko A, Trybala A, Kovalchuk N, Starov V. Current applications of foams formed from mixed surfactant–polymer solutions. Advances in Colloid and Interface Science. 2015;**222**:670-677. DOI: 10.1016/j.cis.2014.10.001

[21] Hajimohammadi R, Johari-Ahar S. Synergistic effect of saponin and rhamnolipid biosurfactants systems on foam behavior. Tenside Surfactants Detergents. 2018;**55**:121-126. DOI: 10.3139/113.110546

[22] Ross J, Miles GD. An apparatus for comparison of foaming properties of soaps and detergents. Oil & Soap. 1941;**18**:99-102. DOI: 10.1007/BF02545418

[23] Li M, Yang W, Chen Z, Qian J, Wang C, Fu S. Phase behavior and polymerization of lyotropic phases. II. A series of polymerizable amphiphiles with systematically varied critical packing parameters. Journal of Polymer Science Part A: Polymer Chemistry. 2006;**44**:5887-5897. DOI: 10.1002/pola.21573

[24] Muntaha ST, Khan MN. Natural surfactant extracted from *Sapindus mukurossi* as an eco-friendly alternate to synthetic surfactant—a dye surfactant interaction study. Journal of Cleaner Production. 2015;**93**:145-150. DOI: 10.1016/j.jclepro.2015.01.023

[25] Zhou C, Wang D, Cao M, Chen Y, Liu Z, Wu C, et al. Self-aggregation, antibacterial activity, and mildness of cyclodextrin/cationic trimeric surfactant complexes. ACS Applied Materials & Interfaces. 2016;**8**:30811-30823. DOI: 10.1021/acsami.6b11667

[26] Lips A, Ananthapadmanabhan K, Vethamuthu M, Hua X, Yang L, Vincent C, et al. Role of surfactant micelle charge in protein denaturation and surfactant-induced skin irritation. In: Surfactants in Personal Care Products and Decorative Cosmetics. Florida, United States: CRC Press; 2006. pp. 184-194. DOI: 10.1201/9781420016123.ch9

Perspective Chapter: Microemulsion as a Game Changer to Conquer Cancer with an Emphasis on Herbal Compounds

S.K. Janani, Raman Sureshkumar and S.P. Dhanabal

Abstract

Microemulsions are lipid based drug delivery system consisting of oil, water, surfactant and often a co-surfactant. They are prepared in order to deliver the drug in an effective manner so as to obtain the desired therapeutic activity. Compared to other conventional therapy, they can deliver the drug in an efficient manner because of their characteristics like reduced particles size, lipid based drug delivery system, thermodynamic stability and economical scale up. Anti-cancer drugs can be easily incorporated into microemulsion so as to target the cancer cells. This helps in increasing the solubility, permeability and absorption of the poorly soluble and poorly permeable drugs, thereby helping in enhancing the bioavailability of the drug. In this chapter, we are also focusing on the herbal based formulations that will be helpful in effectively fighting against cancer cells with less or no side effects. A light has also been shed on the advantages and disadvantages of the microemulsions that will be helpful in considering them as an effective model to conquer cancer and promote the same in the upcoming years.

Keywords: cancer, microemulsion, anti-proliferative activity, herbal compound, lipid based delivery

1. Introduction

Microemulsions are lipid based drug delivery system consisting of oil, water, surfactant and often a co-surfactant. They are prepared in order to deliver the drug in an effective manner so as to obtain the desired therapeutic activity [1]. Compared to other conventional therapy, they can deliver the drug in an efficient manner because of their characteristics like reduced particles size, lipid based drug delivery system, thermodynamic stability and economical scale up [2]. Anti-cancer drugs can be easily incorporated into microemulsion so as to target the cancer cells [3]. In this chapter, we are also focusing on the herbal based formulations; this will be helpful in effectively fighting against cancer cells with less or no side effects. The advantages and limitation of the microemulsions discussed in this chapter will also be helpful in considering them as an effective model to conquer cancer and promote the same in the upcoming years.

As said earlier, the microemulsions are lipid based drug delivery systems that are involved in effectively delivering the drug to the target site [4, 5]. The term microemulsions were employed by T.P. Hoar and J.H. Shulman in 1943 [6]. The word microemulsion was also called as transparent emulsion, solubilized oil

Sl. No.	Water in oil microemulsion	Oil in water microemulsion	Bicontinuous microemulsion
1.			
2.	W/O Microemulsion in which the hydrophilic head is facing inward and the hydrophobic tail is arranged outwards	O/W Microemulsion in which the hydrophilic head is facing outward and the hydrophobic tail is arranged inwards	Bicontinuous microemulsion that has high solubilizing capacity
	-Hydrophilic head	-Hydrophobic tail	

Table 1.
Representing different types of microemulsions.

and micellar solution etc. The microemulsions are optically isotropic in nature and they are also thermodynamically stable. It has been stated that according to IUPAC the diameter of the particle can approximately vary from 1 to 100 nm that can be usually between 10 and 50 nm [7]. Basically the microemulsions can be divided into oil in water (O/W) type, water in oil (W/O) type and bicontinuous microemulsion with high solubilizing power [8, 9]. The different types of micro-emulsion are given in **Table 1**.

While coming to the theories of microemulsion, there are basically three theories as follows:

1. Interfacial theory: Also called as mixed film theory or dual film theory. In this the microemulsion are formed spontaneously by a complex film formation at the oil water interface by the surfactant and co-surfactant. These two com-pounds are further helpful in decreasing the interfacial tension. Thus, forming a microemulsion [10].

2. Solubilization theory: They are usually depicted by the phase diagram by con-sidering components of microemulsion including oil, water and surfactant [11].

3. Thermodynamic theory: The thermodynamic properties of microemulsion like free energy, interfacial tension and surface tension are interdependent on each other. It states that spontaneous emulsification is involved in the formation of microemulsion with negative free energy that makes the emulsion thermody-namically stable [6, 12].

2. Composition and method of preparation of microemulsion

2.1 Composition of microemulsions

Basically, the mircoemulsion consists of three main ingredients including water, oil and surfactant.

2.1.1 Water phase

Aqueous phase plays an important role in the formulation of microemulsion. Based on the type of microemulsion, they act as either a dispersion medium or a dispersed phase in the microemulsion [13]. This phase can accommodate the hydro-philic active drug and some of the preservatives. Sometimes they are also replaced by the buffer solution. In case of water in oil microemulsion, the addition of the water makes changes in the concentration of the surfactant/ water ratio. Dilution of the microemulsion further with the water leads to the phase separation thus, disrupting the droplet formation. Hence, in order to obtain a stable microemulsion, the accurate water ratio must be employed [6].

2.1.2 Oil phase

They are the most important component in the microemulsion system to solu-bilize and transport the lipophilic compounds via the lymphatic system. It also increases the GI (gastrointestinal) absorption of the drug. The selection of oil is determined based on the solubility of the drug in that particular oil. Example of oil includes olive oil, castor oil, capryol 90, oleic acid and isopropyl myristate etc. [6, 8, 14].

2.1.3 Surfactants

They are usually employed in order to decrease the surface tension as well as the interfacial tension. Since, they contain both the lipophilic as well as hydrophilic character; they have the affinity towards both the polar and the non-polar phase. When the aqueous phase is more compared to the oily phase, then the interface between the oil and aqueous phase will curve spontaneously towards the water leading to the formation of oil in water structure. Whereas, when the lipophilic group is more bulky than that of aqueous phase, then the interface will curve towards the opposite direction leading to water in oil microemulsion [15]. In brief the different types of surfactants used in microemulsion are given in **Table 2**. While coming to the type of surfactant, they can be divided into the following:

i. Non-ionic surfactant: This type of surfactant produces less irritation. Thus, they are useful in oral administration. Example includes cremophor, polyoxyethylene sorbitol hexaoleate [16], and tween 20 [17].

ii. Cationic surfactant: Their surface is positively charged. But, as compared to the anionic surfactant, they are more irritating. Example Cetalkonium chloride and benzalkonium chloride etc.

iii. Anionic surfactant: They are negatively charged and possess the ability to penetrate into the skin. Example includes sulphonates, sulphates, and phosphates [8].

iv. Zwitterionic: They are otherwise called as amphoteric surfactant, which contains both positive as well as negative charge that contributes to their neutral charge at neutral pH [18]. Example includes betaines and amphoacetate class compounds.

2.2 Preparation of microemulsions

Microemulsions can be prepared by the following methods:

1. Phase titration method: In this method, the microemulsion is formed by the spontaneous emulsification method. The phase diagram is useful in determining the various interactions that happens when the components of the microemulsion

Sl. No.	Type of surfactants	
1.		Non-ionic surfactant
2.		Cationic surfactant
3.		Anionic surfactant
4.		Zwitterionic surfactant

Table 2.
Indicating the types of surfactants.

is being mixed [19]. Mainly from the phase diagram, the microemulsion region can be determined [20]. Later, the optimized formulation can be obtained by implementing any of the experimental models [21, 22].

2. Phase inversion method: While coming to the phase inversion method, the microemulsion undergoes phase inversion on addition of excess dispersed phase. During this, change in the particle size occurs that can affect the drug release both *in-vitro* as well as *in-vivo* [23]. In this case, when the non-ionic surfactant is used for the preparation of microemulsion, the change in temperature decides the formation of the microemulsion by the method called phase inversion temperature (PIT) method. Suppose when there is an increase in temperature, it leads to the water in oil microemulsion, whereas when the temperature decreases it leads to oil in water microemulsion [24]. While coming to the phase inversion composition, the addition of the water or oil makes the difference [25]. In which, when water is added to oil phase or oil is added to water phase, the surfactant exhibits both the hydrophilic and lipophilic property at one point, and that point is called as emulsion inversion point. After which an addition of water or oil makes the changes in the curvature of the microemulsion and leads to the formation of o/w microemulion or w/o microemulsion [26–28].

3. Herbal based microemulsion for cancer treatment

Various herbal based components are being tested by researchers for determining the anti-cancer activity. But some of the components have low bioavailability because of low solubility. Thus, the microemulsion acts as a carrier for the delivery of such components and helps in improved bioavailability and resulting in enhanced therapeutic effect [29].

Dibenzoylmethane (DBM) is one such compound that has poor water solubility because of which they have low bioavailability. In order to overcome this condition, DBM has been incorporated into microemulsion [30]. The microemulsions are also capable of protecting the drug molecule form enzymatic hydrolysis and oxidation etc., thus allowing the drug to reach the target site. In the study conducted with DBM, peppermint oil has been used and oil in water microemulsion has been made because of poor solubility of DBM in water. This is one of the advantage of microemulsion were both the hydrophilic as well as hydrophobic drug can be incorporated into them. This formulation has exhibited its action as effective chemopreventive agent in forestomach tumor. It has also been stated that the microemulsion decreases the interfacial tension between the vehicle and the intestinal cells and leads to enhanced permeation. This can eventually result in increased therapeutic activity [31].

There are also many medicinal values present in the food compounds that we consume in our day to day life. One such compound is turmeric. The turmeric which is also known as *Curcuma longa L.* (*C. longa*) possesses various biological activities like anti-cancer, anti-oxidant and anti-inflammation etc. [32]. Various studies have been carried out with curcumin on different cancer cells like melanoma, prostate, hepatoma and breast cancer cells. Yen Chu Chen et al. has worked on curcuminoid microemulsion to improve the inhibitory effect on the colon cancer cells. The microemulsion has been prepared by soya bean oil, anhydrous ethanol (co-solvent) and tween 80. Both the early and late apoptosis has been observed in cancer cells and also the cell cycle arrest at S phase. Additionally, the up regulation in p53 (tumor suppressor) has been observed in the p21- independent manner. Increase in caspase 8, caspase 9, and caspase 3 have also been observed. Thus, it can be said that curcumin can produce a wide range of anticancer activity by focusing on extensive mechanism [33].

As said earlier even though various natural compounds are available for the rehabilitation of cancer, they fail to produce the complete desired action because of low bioavailability and poor solubility. Thus, microemulsion helps in enhancing the bioavailability and improving the solubility with the aid of their lipid based drug delivery system. One such study has been carried out by Salma A. El-Sawi et al., to improve the delivery of the extracts of *Salix mucronata* that possesses anti-proliferative activity. The extracts from the *Salix mucronata* has been loaded into the microemulsion, so as to improve the penetration of compounds through the biological barriers and produce a desired pharmacological action with less quantity of dose [34].

When a drug is having low solubility in water but lipophilic in nature, it will be loaded into the microemulsion because of its lipid based delivery. Additionally, because of the lipid based drug delivery, they not only improve the solubility, it can also enhance the permeability of the drugs through various barriers. In which the blood brain barrier (BBB), one of the strongest barriers which do not allow the entry of drugs into the target site, acts as a main hurdle for various drugs in targeting the brain tumor [35]. Even though various flavonoids, terpenes, and carotinoids are being studied to determine their anti-cancer activity, most of them fail to reach the target site because of BBB. Thus, it is important to breach the BBB to get a therapeutic activity. Most of the techniques for delivery of the drug into the brain by crossing the BBB involves invasive or a semi-invasive technique, whereas the nasal route of drug delivery acts as a non-invasive method. But the direct administration of active ingredient into the nasal cavity may lead to degradation of drug by various enzymes and clearance of drug. Thus, in order to overcome these drawbacks, the microemulsion can be used as carrier because of their property to improve the permeability of the drug [36, 37].

Quercetin is a flavonoid that has been widely studied for their anti-cancer activity [38]. They have the ability to inhibit the proliferation of various cancer cells like lung, prostate, liver, breast, cervical and colon cancer [39]. Quercetin is also being used as a synergistic agent along with other chemotherapeutic drugs [40]. But because of their low permeability, they are unable to reach the target site and produce a complete therapeutic activity. Thus, if they are being loaded by any of the carrier or drug delivery system, one can meet the expected outcome. Sagar Kishor Savale in 2017, has prepared the quercetin microemulsion, in order to target the brain tumor assuming that this carrier would help in increasing the permeability of the drug. Quercetin has been dissolved in oleic acid and later tween 80 and polyethylene glycol 400 has been used as surfactant and co-surfactant respectively for the preparation of microemulsion. Though the *in-vitro* permeability study has been done, the in-depth molecular mechanism and *in-vivo* studies should be determined [41].

Another flavonoid that is helpful in targeting the cancer cells is myricetin. It is also structurally similar to that of quercetin and luteolin. Myricetin exhibits anti-cancer activity [42] by interacting with various oncoproteins like Janus kinase – signal transducer and activator of transcription 3 (JAK1-STAT3), protein kinase B (PKB) and mitogen activated kinase 1 (MEK 1). They also act on the over expressed cyclin- dependent kinase 1 to exert an anti-mitotic effect on the liver cancer cells [43]. Though, they exhibit various anticancer activities, the main struggle comes when there is low bioavailability. But this problem can be eradicated by using microemulsion. Studies are also being carried out for the preparation myricetin microemulsion that can improve the bioavailability of the compound [44]. Shujun Wang et al. have formulated myricetin microemulsion for oral delivery. It was stated that the incorporation of microemulsion delivery system could decrease the dose of the myricetin because improved drug availability at the target site. Since, the microemulsion is involved; the author says that the enhanced bioavailability of myricetin at the target site may because of the presence of payers patch and the M (Microfold) cell in the intestine and also because of the presence of lipid in the microemulsion, which makes the enhanced absorption via the lymphatic system [45].

Ruixue Guo et al. has prepared a myricetin microemulsion, for enhancing the bioavailability of the compound, anti-proliferative activity and also the anti-oxidant activity. In this study, the effect of myricetin microemulsion was tested on the liver cancer cells (HepG2). Since, the excess addition of surfactant in the microemulsion may lead to toxicity, it is important that we need to use the surfactant in the minimum quantity. Thus focusing on these criteria, the work has been carried out by screening various oils, surfactant and co-surfactant to get stable and optimized formulation. As a result of it, labrafac lipophile WL 1349 was selected as oil phase, tween 80 and cremophor RH 40 was used as surfactant and transcutol HP has been used a co-surfactant. The prepared miroemulsion has shown enhanced oral bioavailability as compared to that of the myricetin suspension (Sodium carboxy methylcellulose suspension) [46].

The microemulsion can also be a self-micro-emulsifying drug delivery system (SMEDDS). They are given as a preconcentrate so that after reaching the target site, they may get converted into a microemulsion giving a prodrug effect. The advantage of such delivery system is that, the stability of the formulation can be maintained as a result the bioavailability can also be improved. In the SMEDDS, the lipid phase used can also be liquid or solid. While comparing both, it has been said that S-SMEDDS (i.e.): use of solid lipid is preferred because, it involves the solidification of liquid excipients into a powder form (solid form) that helps in obtaining a more stable formulation compared to liquid lipid. One such solid SMEDDS has been prepared by Wenli Huang et al. [47]. In this work, *Brucea javanica* oil is being loaded into S-SMEDDS (BJOS) to enhance the anti-cancer activity. *Brucea javanica* oil is a traditional herbal medicine in China that has the ability to treat various cancers like gastrointestinal cancer, lung cancer and prostate cancer. Since, the oil itself is producing various anti-cancer activity, they were used as a main component in the study that can act as the anti-cancer agent. The BJOS formulation was prepared by using *Brucea javanica* as oil phase, Cremophor RH40 and PEG 6000 as surfactant and co-surfactant respectively. Compared to the Brucea javanica oil emulsion alone, the BJOS formulation had a better anti-proliferative activity. This may be because; the S-SMEDDS would have carried the BJO into the cancer cells via endocytosis.

Apart from the above said compounds, there are also other herbal compounds that exhibits anti-cancer activity by acting on various pathways by inhibiting the cell cycle arrest, maintaining the genetic stability, inducing apoptosis exhibiting anti-angiogenic activity and also anti-metastatic activity. Thus, as a nutshell, it can be said that it would be a better option if we choose microemulsion as a drug delivery system or a carrier for the delivery of herbal compounds that exhibits anti-cancer activity. Because unless the conventionally available drugs or synthetic drugs, the herbal compounds may not cause much side effects or harm the normal cells. Even the herbal compounds are also being taken by us on regular basis. Thus, it will lead to great endeavors if natural compounds are being delivered in an accurate manner by using this kind of delivery system.

4. Different routes of administration of microemulsion

4.1 Oral route

As it was already said that the solubility of the drug plays an important role in bioavailability of the drug, it is important to improve the solubility so as to increase the bioavailability (**Figure 1**). When a drug is being incorporated in microemulsion, compared to conventional medications, the microemulsion can increase the absorption, increase the therapeutic efficacy and also decrease the drug toxicity [12]. Apart from the solubility, in many of the cancers, the MDR (multi drug resistance) plays an important role; it does not allow the anti-cancer drugs to produce its activity by creating an efflux system. But, the microemulsion has the capacity to overcome this efflux system and deliver the anti-cancer drug in a more efficient manner. One such study has been carried out by Ding Qu et al., in which a multicomponent microemulion has been formulated consisting of coix seed oil, ginsenoside Rh2 loaded with etoposide. This formulation has been useful in inhibiting the Pgp-efflux, which may because of the use of G-Rh2 that has the capacity to interact with the Pgp-efflux and in addition, the formulation has produced a synergistic activity which may be because of the oils used in this formulation, that also possess some anti-cancer activity [48].

Apart from the normal microemulsion, the microemulsion can be used as a preconcentrate in order to improve the bioavailability as well as therapeutic effect of anti-cancer drug. As it was said already, one such form of preconcentrate is called the Self-microemulsifying drug delivery system (SMEDDS). They are also similar to

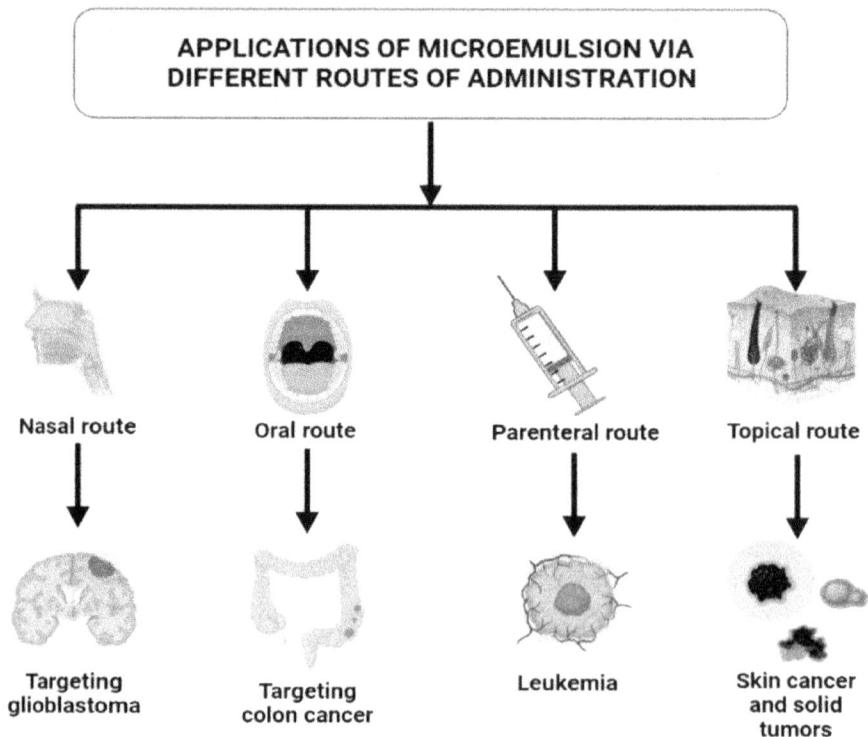

APPLICATIONS OF MICROEMULSION VIA
DIFFERENT ROUTES OF ADMINISTRATION

Nasal route	Oral route	Parenteral route	Topical route
Targeting glioblastoma	Targeting colon cancer	Leukemia	Skin cancer and solid tumors

Figure 1.
Represents the different routes of administration of microemulsion for targeting various cancers (Created with Biorender).

that of micro-emulsion, whereas being a preconcentrate, they does not have water in their formulation. During this case a question arises in our mind stating that were does the aqueous phase come from. Well, the SMEDDS being a preconcetrae when administered, it gets emulsified with the help of Gastro intestinal fluid, thereby forming a microemulsion. This type of self-emulsifying microemulsion serves more advantage as compared to the existing microemulsion. Since, the water is not involved, enhanced physical and chemical stability can be obtained on long term storage [49]. It also has the ability to decrease the gastric irritation caused by some of the anti-cancer agents. Studies are also being carried out with SMEDDS for the rehabilitation of cancer. Triptolide obtained from *Tripterygium wilfordii* possess anti-cancer activity, at the same time they also have low solubility, gastrointestinal irritation and also poor bioavailability. Triptolide has been incorporated to SMEDDS in order to overcome the above said effects and their anticancer activity on gastric cancer has also been identified [50].

4.2 Parentral route

Microemulsion can be delivered via the parenteral route instead of suspension that is not suitable for the parenteral route. Compared to liposomes, they are having a good stability when given through parenteral route. When a microemulsion is loaded with the higher concentration of the drug, the frequency of administration can be decreased [51]. Some of the drugs have to be given through the parenteral route because in order to overcome the degradation of the drug, instability of the drug, low bioavailability of the drug, decreased therapeutic activity of the drug due to oral administration. At some instance, the parenteral route of administration of the drug itself can lead to precipitation of the drug when mixed with the infusion fluids. Thus, in those cases it is important to incorporate the drug into a suitable vehicles or carrier to obtain an improved bioavailability an increased therapeutic activity without any capillary blockade. But incorporation of more solvents or aqueous solutions may cause mild to severe side effects. In these circumstances, the microemulsion comes into play, because of their thermodynamic stability. Jayesh Jain et al. has worked on etoposide, in which they have formulated the microemulsion for parenteral administration. Since, the etoposide need to be administered slowly through the intravenous infusion, it is important to detect the changes when they are mixed with infusion fluids. As a result they observed that the microemulsion was not having a drug precipitation up to a certain concentration of drug. But as the concentration of drug increases the precipitation serves as a limitation. Thus, more study on the reason for the drug precipitation should be understood [52].

The microemulsion also serve various advantages while comparing other colloidal carriers, microemulsions can deliver the hydrophobic drugs, as they can solubilize those types of drugs and also the scale-up can be done easily compared to other carriers.

It is important to know the right excipients for the preparation of microemulsions for parenteral delivery, as the parenteral delivery deal with only few excipients. Those excipients should also be biocompatible, non- irritant and a sterilizable one. Other than these properties, the excipients that are used for preparation of microemulsion for parenteral delivery is similar to that of the normal microemulsion. They consist of oily phase, surfactant, co-surfactant and the aqueous phase. But while using these components during the formulation, certain factors should be considered like [53]:

- Long term usage of oily phase for parenteral administration should be determined.

- Certain surfactants like polyoxyethylene alky ethers may cause hemolysis at higher concentrations. So, the concentration of this type of surfactants must be considered.

- The use of polyethoxylated castor oil which has to be used with caution because of their ability to cause serious adverse effects.

- The use of propylene glycol as co-surfactant at higher concentrations my cause pain at the injection site and also hemolysis.

4.3 Nasal route

Glioblastoma, or brain tumor, is the condition, which is very difficult to target by various anti-cancer drugs. The administration of drug through oral route may not guarantee that it will deliver most of drug to the target site. And also most of the available techniques for targeting brain tumors are either invasive or semi-invasive. Thus, it is important to deliver a drug in such a way that it can reach the target site without being invasive in nature and also improve the bioavailability with enhanced therapeutic activity. Nasal route of administration comes into play at these situations. It is one of the non-invasive techniques that can bypass the BBB. When a drug is being administered intranasally, they can enter into the brain or target site depending upon the nature of the drug, type of formulation and the physiological conditions. The different pathways of entry of drug are [54]:

- Olfactory nerve

- Trigeminal nerve

- Lymphatic pathway

- Cerebro spinal fluid (CSF)

- Vascular pathway.

Various flavonoids like curcumin [55] and rhein [56] are also been administered via intranasal delivery to study the effect of these compounds on glioblastoma, as they already possess various anti-cancer activity, anti-oxidant activity and anti-inflammatory etc.

It has also mentioned in previous studies that the uses of microemulsion for the intranasal delivery are safe and effective. And moreover, since the nasal route helps in direct delivery of drug to the target site than entering into systemic circulation, the concentration of drug at the target site can be increased along with the use of microemulsion [57]. The concentration of various drugs have been improved at the target site with the help of microemulsion and some of them include albendazole sulfoxide that possess anti-proliferative activity [58] and teriflunomide that also possess anti-cancer activity [59–61].

The administration of drug through nasal route may be prone to mucociliary clearance. Thus, in this case the drug or the formulation may not be able to reach the target site and produce its therapeutic action. Thus, for this purpose, the mucoadhesive agents are being used that can slow down the mucociliary clearance [62]. Because of the enhanced penetration property of microemulsions through biological membranes, they are being used to deliver the drug through nasal cavity [63]. Thus, the use of mucoadhesive agents to the microemulsion can improve the retention time and increase the absorption of drug. Julio Mena-Hernández et al. have formulated mebendazole microemulsion containing mucoadhesive agent namely sodium hyaluronate for intranasal delivery to target the glioblastoma. It was stated that the treatment with the formulation has increased the survival time in animals and also decreased nasal toxicity has been observed [64].

4.4 Topical route

Topical route is one of the alternative approaches over the oral and parenteral route. This is one of the non-invasive methods of drug delivery system [65]. They are also convenient route of drug delivery. They are also helpful in overcoming various limitations that are being produced by oral route like gastrointestinal degradation of drug, hepatic clearance, toxicity and finally decreased bioavailability. Though the topical route have several advantages, they also have certain disadvantages like low permeability though the skin, decreased residence time on skin, the viscosity of the formulation, spreadability of the formulation etc. But these problems can be conquered with the aid of microemulsion. They have the capacity to permeate through the skin, improve the solubility of the drug and also enhance the absorption of the drug. Various drugs are being incorporated into the microemulsion because of these properties [66]. Example Methyl dihydrojasmonate that has the capacity to produce anti-tumor activity is being incorporated into microemulsion and studied. Targeting solid tumors by transdermal delivery is in emerging stage. This allows convenient way of targeting the tumor and also overcome the first pass metabolism produced by the oral dosage forms. The Methyl dihydrojasmonate that has been incorporated in microemulsion has been studied on MCF-7 cancer cell and Ehrlich solid carcinoma model [67].

The non-toxic and non-irritant property of microemulsion makes them suitable to treat skin cancers. 5- Fluorouracil has been used for treating skin cancer. But they have a major disadvantage of poor skin permeation. Thus, in this case they are being incorporated into microemulsion for improving the permeability of the drug. So that it can produce its anti-cancer activity [68]. 5-fluorouracil is being studied by many of the researchers by incorporating them into a microemulsion for rehabilitation of skin cancer [69]. Thus, opting for a microemulsion would be a good strategy for the treatment of skin cancer and solid tumors.

Apart from the microemulsion, the incorporation of microemulsion into gel also plays a vital role. It further increases the residence time of the drug on the skin and also release of the drug on the target site is prolonged [70].

Sl. No.	Drug used	Type of cancer	Oil	Surfactant	Co-Surfactant	Reference
1.	Imiquimod	Colon cancer	Coconut oil	Lecithin	Tween 60	[71]
2.	Simvastatin	Colon and liver cancer	Captex	Cremophor EL	Transcutol	[72]
3.	Methyl dihydrojasmonate	Ehrlich tumor cells	Pure soybean oil	Coconut salt fatty acids and Phosphatidylcholine	Glycerol	[73]
4.	βelemene and celastrol	Lung cancer	Labrafil	Kolliphor	Polyethylene glycol 400	[74]
5.	5-Fluorouracil	Skin tumor	Isopropyl myristate	Tween 80	Span 20	[75]

Table 3.
Various microemulsions with different drug, oil, surfactant and co-surfactant are given.

Some other microemulsions that are studied by various researchers for anti-cancer activity is given in **Table 3**.

5. Advantages and disadvantages of microemulsion

5.1 Advantages of microemulsion

- Due to the amphiphilic nature, they can solubilize both hydrophobic and hydrophilic drugs [76].

- They are thermodynamically stable.

- Can deliver the maximum amount of drug to the target site. Thus, helps in decreasing the side effects of the drug.

- They also have long shelf life.

- Scale –up is easy and requires less energy.

- Improved lymphatic delivery of the drugs due to the usage of lipid phase.

- Good permeability can be obtained for low permeable drugs.

- Enhanced solubility and thereby improved bioavailability.

- Absorption can also be increased.

- Can overcome the first pass metabolism [77].

- Can mask the unpleasant taste.

- Can be administered through various routes like oral, parenteral, topical and intranasal.

- Increase in patient compliance due to liquid dosage form.

5.2 Disadvantages of microemulsion

- Large quantity of surfactant and co-surfactant usage may lead to toxicity issues [76].

- It's a doubt when it comes to sustained drug release [78].

- Due to the precipitation of the drug, they should be used in infusion or parenterals after careful investigation of nature of the drug and the microemulsion character.

- Since, there is a toxicity issue with surfactant, the surfactant should fall under the "Generally-Regarded-as-Safe" (GRAS) class [79].

6. Future perspectives

Microemulsions are the lipid based drug delivery systems that are being studied by a wide group of researchers. Even though the microemulsion has various advantages like loading of both hydrophilic and lipophilic drug, thermodynamic stability and improved patient compliance etc. They lack behind when it comes to the surfactant concentration. The main ingredient used to decrease the interfacial tension between the oil phase and aqueous phase is surfactant and the co-surfactant. Being the main ingredient, they may be used in large quantity. This turns out to be one of the major disadvantages, which thereby leads to toxicity. Usually only 40% of the surfactant should be used in the formulation. Thus, a suitable surfactant that is capable of decreasing the interfacial tension at a low concentration must be employed in the formulation [80]. The microemulsions are usually suitable for scale up process, which is more important from industrial point of view that can be done at a low cost. Apart from the treating various cancers and diseases, the microemulsions are also being used for the cosmetic formulations [81], because of their increased permeability and action on the skin for a prolonged period of time. Though they are being used for various route of delivery for treating a wide range of cancer, the surfactant concentration should be considered which plays a major role in deciding the fate of microemulsion.

7. Conclusions

Microemulsions are lipid based drug delivery system that comprises of oil, surfactant and a co-surfactant. This can solubilize both the hydrophilic as well as lipophilic drugs. Thus, helps in increasing the absorption and bioavailability of the drugs. Since, the bioavailability is increased; the desired therapeutic activity can be obtained with minimum dose of the drug. Thu, we can also overcome the toxicity due to higher dose of the drug. Most of the herbal compounds like flavonoids, terpenes and caratinoids are possessing the anti-cancer activity. But only because of their poor solubility and poor permeability, they are not being widely studied for conquering cancer cells. This fate of herbal compounds can be changed once they are being incorporated into a safe and effective carrier or vehicle. Even though various carrier systems are available and being studied widely by researchers, the microemulsion plays a vital role and standalone compared to others because of their property like thermodynamic stability, increasing the solubility and permeability of drug and also easy scale up process. They act as a stalwart supporter in the delivery of potential anti-cancer compounds, without affecting the nature of the drug and maintaining their stability by overcoming various hurdles like GI (Gastrointestinal) degradation, first pass metabolism and also toxicity due to higher dose of the drug. Thus, as a nutshell it can be said that the microemulsions are prominent carriers in delivering the effective anti-cancer drug and also responsible for increasing the patient compliance and enhancing the therapeutic activity. But, apart from this, the limitations of the carrier should also be taken into consideration, like use of desired oil, surfactant and co-surfactant as well as their concentrations. By meticulously pondering the above points, one can develop an effective formulation for the rehabilitation of the cancer with the aid of microemulsions.

Acknowledgements

The authors would like to thank Indian Council of Medical Research (ICMR), New Delhi for awarding Senior Research Fellowship (SRF)-"3/ 2/2/2/2020-NCD-III."

The authors would like to thank the Department of Science and Technology – Fund for Improvement of Science and Technology Infrastructure in Universities and Higher Educational Institutions (DST-FIST), New Delhi for their infrastructure support to our department.

Conflict of interest

The authors declare no conflict of interest.

Author details

S.K. Janani[1], Raman Sureshkumar[1*] and S.P. Dhanabal[2]

1 Department of Pharmaceutics, JSS College of Pharmacy, JSS Academy of Higher Education and Research, Ooty, Nilgiris, Tamil Nadu, India

2 Department of Pharmacognosy and Phytopharmacy, JSS College of Pharmacy, JSS Academy of Higher Education and Research, Ooty, Nilgiris, Tamil Nadu, India

*Address all correspondence to: sureshcoonoor@jssuni.edu.in

IntechOpen

References

[1] He CX, He ZG, Gao JQ. Microemulsions as drug delivery systems to improve the solubility and the bioavailability of poorly water-soluble drugs. Expert Opinion on Drug Delivery. 2010;**7**(4):445-460. DOI: 10.1517/17425241003596337

[2] Mahdi ZH, Maraie NK. Overview on Nanoemulsion as a recently developed approach in Drug Nanoformulation. Research Journal of Pharmacy and Technology. 2019;**12**(11):5554-5560

[3] Fanun M. Microemulsions as delivery systems. Current Opinion in Colloid & Interface Science. 2012;**17**(5):306-313. DOI: 10.1016/j.cocis.2012.06.001

[4] Cannon JB, Long MA. Emulsions, microemulsions, and lipid-based drug delivery systems for drug solubilization and delivery—Part II: Oral applications. Water-Insoluble Drug Formulation. 2018;**12**:247-282

[5] Ghosh PK, Murthy RS. Microemulsions: A potential drug delivery system. Current Drug Delivery. 2006;**3**(2):167-180. DOI: 10.2174/156720106776359168

[6] Sharma AK, Garg T, Goyal AK, Rath G. Role of microemuslsions in advanced drug delivery. Artificial Cells, Nanomedicine, and Biotechnology. 2016;**44**(4):1177-1185. DOI: 10.3109/21691401.2015.1012261

[7] IUPAC. Definition [Internet]. 2021. Available from: https://surfactantsandmicroemulsions.wordpress.com/microemulsion/iupac-definition/ [Accessed: May 18, 2021]

[8] Rajput R, Kumar V, Sharma S. Microemulsions: Current trends in sustained release drug delivery systems. International Journal of Pharma Professional's Research. 2016;**7**(1):1326-1332

[9] Dantas TNC, Santanna VC, Souza TTC, Lucas CRS, Dantas Neto AA, Aum PTP. Microemulsions and nanoemulsions applied to well stimulation and enhanced oil recovery (EOR). Brazilian Journal of Petroleum and Gas. 2019;**12**(4):251-265. DOI: 10.5419/bjpg2018-0023

[10] GR D. Microemulsions: Platform for improvement of solubility and dissolution of poorly soluble drugs. Asian Journal of Pharmaceutical and Clinical Research. 2015;**8**(5):7-17

[11] Kale SN, Deore SL. Emulsion micro emulsion and nano emulsion: A review. Systematic Reviews in Pharmacy. 2017;**8**(1):39. DOI: 10.5530/srp.2017.1.8

[12] Lawrence MJ, Rees GD. Microemulsion-based media as novel drug delivery systems. Advanced Drug Delivery Reviews. 2000;**45**(1):89-121. DOI: 10.1016/S0169-409X(00)00103-4

[13] Sahu GK, Sharma H, Gupta A, Kaur CD. Advancements in microemulsion based drug delivery systems for better therapeutic effects. International Journal of Pharmaceutical Sciences and Developmental Research. 2015;**1**(1):008-015

[14] Abd Sisak MA, Daik R, Ramli S. Study on the effect of oil phase and co-surfactant on microemulsion systems. Malaysian Journal of Analytical Sciences. 2017;**21**.1409-1416. DOI: 10.17576/mjas-2017-2106-23

[15] Langevin D. Microemulsions. Accounts of Chemical Research. 1988;**21**(7):255-260

[16] Joubran RF, Cornell DG, Parris N. Microemulsions of triglyceride and non-ionic surfactant—Effect of temperature and aqueous phase composition. Colloids and Surfaces A: Physicochemical and Engineering

Aspects. 1993;**80**(2-3):153-160. DOI: 10.1016/0927-7757(93)80194-J

[17] Chen H, Guan Y, Zhong Q. Microemulsions based on a sunflower lecithin–Tween 20 blend have high capacity for dissolving peppermint oil and stabilizing coenzyme Q10. Journal of Agricultural and Food Chemistry. 2015;**63**(3):983-989. DOI: 10.1021/jf504146t

[18] Soleimani Zohr Shiri M, Henderson W, Mucalo MR. A review of the lesser-studied microemulsion-based synthesis methodologies used for preparing nanoparticle systems of the noble metals, Os, Re, Ir and Rh. Materials. 2019;**12**(12):1896. DOI: 10.3390/ma12121896

[19] Mehta DP. Microemulsions: A potential novel drug delivery system. International Journal of Pharmaceutical Sciences. 2015;**1**:48

[20] Anand S, Kumar KR. Design, development and optimisation of carvedilol microemulsion by pseudoternary phase diagram and central composite design. International Journal of Research in Pharmaceutical Sciences. 2020;**11**(4):6619-6632. DOI: 10.26452/ijrps.v11i4.3569

[21] Rao S, Barot T, Rajesh KS, Jha LL. Formulation, optimization and evaluation of microemulsion based gel of butenafine hydrochloride for topical delivery by using simplex lattice mixture design. Journal of Pharmaceutical Investigation. 2016;**46**(1):1-2. DOI: 10.1007/s40005-015-0207-y

[22] Juškaitė V, Ramanauskienė K, Briedis V. Design and formulation of optimized microemulsions for dermal delivery of resveratrol. Evidence-based Complementary and Alternative Medicine. 2015;**2015**:1-10. DOI: 10.1155/2015/540916

[23] Saini JK, Nautiyal U, Kumar M, Singh D, Anwar F. Microemulsions: A potential novel drug delivery system. International Journal of Pharmaceutical and Medicinal Research. 2014;**2**(1):15-20

[24] Singh PK, Iqubal MK, Shukla VK, Shuaib M. Microemulsions: Current trends in novel drug delivery systems. Journal of Pharmaceutical, Chemical and Biological Sciences. 2014;**1**(1):39-51

[25] Gupta PK, Bhandari N, Shah H, Khanchandani V, Keerthana R, Nagarajan V, et al. An update on nanoemulsions using nanosized liquid in liquid colloidal systems. Nanoemulsions-Properties, Fabrications and Applications. 2019:1-20. DOI: 10.5772/intechopen.84442

[26] Kaundal A, Choudhary A, Sharma D. Microemulsions: A novel drug delivery system. International Journal of Pharmaceutical Research and Bioscience. 2016;**5**(3):193-210

[27] Lokhande SS. Microemulsions as promising delivery systems: A review. Asian Journal of Pharmaceutical Research. 2019;**9**(2):90-96. DOI: 10.5958/2231-5691.2019.00015.7

[28] Perazzo A, Preziosi V, Guido S. Phase inversion emulsification: Current understanding and applications. Advances in Colloid and Interface Science. 2015;**222**:581-599. DOI: 10.1016/j.cis.2015.01.001

[29] Kawakami K, Yoshikawa T, Hayashi T, Nishihara Y, Masuda K. Microemulsion formulation for enhanced absorption of poorly soluble drugs: II. In vivo study. Journal of Controlled Release. 2002;**81**(1-2):75-82. DOI: 10.1016/S0168-3659(02)00050-0

[30] Puri A, Kaur A, Raza K, Goindi S, Katare OP. Development and evaluation of topical microemulsion of dibenzoylmethane for treatment of UV induced photoaging. Journal of Drug Delivery Science and Technology.

2017;**37**:1-2. DOI: 10.1016/j. jddst.2016.09.010

[31] Kaur A, Sharma G, Verma S, Goindi S, Katare OP. Oral microemulsion of phytoconstituent found in licorice as chemopreventive against benzo (a) pyrene induced forestomach tumors in experimental mice model. Journal of Drug Delivery Science and Technology. 2017;**39**:523-530. DOI: 10.1016/j.jddst.2017.05.006

[32] Rajkumari S, Sanatombi K. Nutritional value, phytochemical composition, and biological activities of edible Curcuma species: A review. International Journal of Food Properties. 2017;**20**(3):S2668-S2687. DOI: 10.1080/10942912.2017.1387556

[33] Chen YC, Chen BH. Preparation of curcuminoid microemulsions from Curcuma longa L. to enhance inhibition effects on growth of colon cancer cells HT-29. RSC Advances. 2018;**8**(5):2323-2337. DOI: 10.1039/C7RA12297G

[34] El-Sawi SA, Maamoun AA, Salama AH, Maamoun MA, Farghaly AA. Application of microemulsion formulation in improving the antiproliferative performance of Salix mucronata (Thunb) leaves with chemical investigation of the active extract. Acta Ecologica Sinica. 2020;**40**(4):339-346. DOI: 10.1016/j. chnaes.2020.04.004

[35] Pandit R, Chen L, Götz J. The blood-brain barrier: Physiology and strategies for drug delivery. Advanced Drug Delivery Reviews. 2020;**165**:1-4. DOI: 10.1016/j.addr.2019.11.009

[36] Froelich A, Osmałek T, Jadach B, Puri V, Michniak-Kohn B. Microemulsion-based media in nose-to-brain drug delivery. Pharmaceutics. 2021;**13**(2):201. DOI: 10.3390/pharmaceutics13020201

[37] Zhang Q, Jiang X, Jiang W, Lu W, Su L, Shi Z. Preparation of nimodipine-loaded microemulsion for intranasal delivery and evaluation on the targeting efficiency to the brain. International Journal of Pharmaceutics. 2004;**275**(1-2): 85-96. DOI: 10.1016/j.ijpharm.2004. 01.039

[38] Ezzati M, Yousefi B, Velaei K, Safa A. A review on anti-cancer properties of Quercetin in breast cancer. Life Sciences. 2020;**248**:117463. DOI: 10.1016/j.lfs.2020.117463

[39] Murakami A, Ashida H, Terao J. Multitargeted cancer prevention by quercetin. Cancer Letters. 2008;**269**(2): 315-325. DOI: 10.1016/j.canlet.2008. 03.046

[40] Rauf A, Imran M, Khan IA, ur-Rehman M, Gilani SA, Mehmood Z, et al. Anticancer potential of quercetin: A comprehensive review. Phytotherapy Research. 2018;**32**(11):2109-2130. DOI: 10.1002/ptr.6155

[41] Savale SK. Formulation and evaluation of quercetin nanoemulsions for treatment of brain tumor via intranasal pathway. Asian Journal of Biomaterial Research. 2017;**3**(6):28-32

[42] Semwal DK, Semwal RB, Combrinck S, Viljoen A. Myricetin: A dietary molecule with diverse biological activities. Nutrients. 2016;**8**(2):90. DOI: 10.3390/nu8020090

[43] Devi KP, Rajavel T, Habtemariam S, Nabavi SF, Nabavi SM. Molecular mechanisms underlying anticancer effects of myricetin. Life Sciences. 2015;**142**:19-25. DOI: 10.1016/j. lfs.2015.10.004

[44] Sakthivel G, Prajisha P, Karunya MD, Ravindran R. Protective effect of myricetin microemulsion against psycho-logical stress in rat model. Journal of Psychiatry and Cognitive Behavior. 2017;**10**:2574-7762. DOI: 10.29011/2574-7762.000022

[45] Wanga S, Yea T, Zhanga X, Yanga R, Yib X. Myricetin microemulsion for oral drug delivery: Formulation optimization, in situ intestinal absorption and in-vivo evaluation. Asian Journal of Pharmaceutical Sciences. 2013;**8**:18-27

[46] Guo RX, Fu X, Chen J, Zhou L, Chen G. Preparation and characterization of microemulsions of myricetin for improving its antiproliferative and antioxidative activities and oral bioavailability. Journal of Agricultural and Food Chemistry. 2016;**64**(32):6286-6294. DOI: 10.1021/acs.jafc.6b02184

[47] Huang W, Su H, Wen L, Shao A, Yang F, Chen G. Enhanced anticancer effect of Brucea javanica oil by solidified self-microemulsifying drug delivery system. Journal of Drug Delivery Science and Technology. 2018;**48**:266-273. DOI: 10.1016/j.jddst.2018.10.001

[48] Qu D, Wang L, Liu M, Shen S, Li T, Liu Y, et al. Oral nanomedicine based on multicomponent microemulsions for drug-resistant breast cancer treatment. Biomacromolecules. 2017;**18**(4):1268-1280. DOI: 10.1021/acs.biomac.7b00011

[49] Dokania S, Joshi AK. Self-microemulsifying drug delivery system (SMEDDS)—Challenges and road ahead. Drug Delivery. 2015;**22**(6):675-690. DOI: 10.3109/10717544.2014.896058

[50] Xie M, Wu J, Ji L, Jiang X, Zhang J, Ge M, et al. Development of triptolide self-microemulsifying drug delivery system and its anti-tumor effect on gastric cancer xenografts. Frontiers in Oncology. 2019;**9**:978. DOI: 10.3389/fonc.2019.00978

[51] Rajpoot K, Tekade RK. Microemulsion as drug and gene delivery vehicle: An inside story. In: Drug Delivery Systems. Cambridge, Massachusetts: Academic Press; 2019. pp. 455-520. DOI: 10.1016/B978-0-12-814487-9.00010-7

[52] Jain J, Fernandes C, Patravale V. Formulation development of parenteral phospholipid-based microemulsion of etoposide. AAPS PharmSciTech. 2010;**11**(2):826-831. DOI: 10.1208/s12249-010-9440-x

[53] Date AA, Nagarsenker MS. Parenteral microemulsions: an overview. International Journal of Pharmaceutics. 2008;**355**(1-2):19-30. DOI: 10.1016/j.ijpharm.2008.01.004

[54] Bruinsmann FA, Richter Vaz G, de Cristo Soares Alves A, Aguirre T, Raffin Pohlmann A, Stanisçuaski Guterres S, et al. Nasal drug delivery of anticancer drugs for the treatment of glioblastoma: Preclinical and clinical trials. Molecules. 2019;**24**(23):4312. DOI: 10.3390/molecules24234312

[55] Mukherjee S, Baidoo J, Fried A, Atwi D, Dolai S, Boockvar J, et al. Curcumin changes the polarity of tumor-associated microglia and eliminates glioblastoma. International Journal of Cancer. 2016;**139**:2838-2849. DOI: 10.1002/ijc.30398

[56] Sun H, Luo G, Chen D, Xiang ZA. Comprehensive and system review for the pharmacological mechanism of action of Rhein, an active anthraquinone ingredient. Frontiers Pharmacology. 2016;7:247. DOI: 10.3389/fphar.2016.00247

[57] Shah B, Khunt D, Misra M, Padh H. Non-invasive intranasal delivery of quetiapine fumarate loaded microemulsion for brain targeting: Formulation, physicochemical and pharmacokinetic consideration. European Journal of Pharmaceutical Sciences. 2016;**91**:196-207. DOI: 10.1016/j.ejps.2016.05.008

[58] Kim U, Shin C, Kim C, Ryu B, Kim J, Bang J, et al. Albendazole exerts antiproliferative effects on prostate cancer cells by inducing reactive oxygen species generation. Oncology Letters.

2021;**21**(5):1-9. DOI: 10.3892/ol.2021.12656

[59] Huang O, Zhang W, Zhi Q, Xue X, Liu H, Shen D, et al. Featured article: Teriflunomide, an immunomodulatory drug, exerts anticancer activity in triple negative breast cancer cells. Experimental Biology and Medicine. 2015;**240**(4): 426-437. DOI: 10.1177/1535370214554881

[60] Shinde RL, Bharkad GP, Devarajan PV. Intranasal microemulsión for targeted nose to brain delivery in neurocysticercosis: Role of docosahexaenoic acid. European Journal of Pharmaceutics and Biopharmaceutics. 2015;**96**(363-79):16. DOI: 10.1016/j.ejpb.2015.08.008

[61] Gadhavea D, Gorainb B, Tagalpallewara A, Kokarea C. Intranasal teriflunomide microemulsion: An improved chemotherapeutic approach in glioblastoma. Journal of Drug Delivery Science and Technology. 2019;**51**:276-289. DOI: 10.1016/j.jddst.2019.02.013

[62] Chaturvedi M, Kumar M, Pathak K. A review on mucoadhesive polymer used in nasal drug delivery system. Journal of Advanced Pharmaceutical Technology & Research. 2011;**2**(4):215. DOI: 10.4103%2F2231-4040.90876

[63] Pathak R, Dash RP, Misra M, Nivsarkar M. Role of mucoadhesive polymers in enhancing delivery of nimodipine microemulsion to brain via intranasal route. Acta Pharmaceutica Sinica B. 2014;**4**(2):151-160. DOI: 10.1016/j.apsb.2014.02.002

[64] Mena-Hernández J, Jung-Cook H, Llaguno-Munive M, García-López P, Ganem-Rondero A, López-Ramírez S, et al. Preparation and evaluation of mebendazole microemulsion for intranasal delivery: An alternative approach for glioblastoma treatment. AAPS PharmSciTech. 2020;**21**(7):1-2. DOI: 10.1208/s12249-020-01805-x

[65] Mathias NR, Hussain MA. Non-invasive systemic drug delivery: Developability considerations for alternate routes of administration. Journal of Pharmaceutical Sciences. 2010;**99**(1):1-20. DOI: 10.1002/jps.21793

[66] Shukla T, Upmanyu N, Agrawal M, Saraf S, Saraf S, Alexander A. Biomedical applications of microemulsion through dermal and transdermal route. Biomedicine & Pharmacotherapy. 2018;**108**:1477-1494. DOI: 10.1016/j.biopha.2018.10.021

[67] Yehia R, Hathout RM, Attia DA, Elmazar MM, Mortada ND. Anti-tumor efficacy of an integrated methyl dihydrojasmonate transdermal microemulsion system targeting breast cancer cells: in vitro and in vivo studies. Colloids and Surfaces B: Biointerfaces. 2017;**155**:512-521. DOI: 10.1016/j.colsurfb.2017.04.031

[68] Sharma H, Sahu GK, Kaur CD. Development of ionic liquid microemulsion for transdermal delivery of a chemotherapeutic agent. SN Applied Sciences. 2021;**3**(2):1-10. DOI: 10.1007/s42452-021-04235-x

[69] Kumar S, Sinha VR. Design, development and characterization of topical microemulsions of 5-fluorouracil for the treatment of non melanoma skin cancer and its precursor lesions. Anti-Cancer Agents in Medicinal Chemistry (Formerly Current Medicinal Chemistry-Anti-Cancer Agents). 2016;**16**(2): 259-268

[70] Scomoroscenco C, Teodorescu M, Raducan A, Stan M, Voicu SN, Trica B, et al. Novel gel microemulsion as topical drug delivery system for curcumin in dermatocosmetics. Pharmaceutics. 2021;**13**(4):505. DOI: 10.3390/pharmaceutics13040505

[71] Rajaram SV, Ravindra PP, Shripal MC. Microemulsion drug delivery of imiquimod as anticancer

agent for skin cancer therapy and its evaluation. Drug Design. 2020;**9**:170

[72] Alkhatib MH, Al-Merabi SS. In vitro assessment of the anticancer activity of simvastatin-loaded microemulsion in liver and colon cancer cells. Journal of Drug Delivery Science and Technology. 2014;**24**(4):373-379. DOI: 10.1016/S1773-2247(14)50076-7

[73] da Silva GB, Scarpa MV, Carlos IZ, Quilles MB, Lia RC, do Egito ES, et al. Oil-in-water biocompatible microemulsion as a carrier for the antitumor drug compound methyl dihydrojasmonate. International Journal of Nanomedicine. 2015;**10**:585. DOI 10.2147%2FIJN.S67652

[74] Zhang Q, Tian X, Cao X. Transferrin-functionalised microemulsion co-delivery of β-elemene and celastrol for enhanced anti-lung cancer treatment and reduced systemic toxicity. Drug Delivery and Translational Research. 2019;**9**(3):667-678. DOI: 10.1007/s13346-019-00623-4

[75] Goindi S, Arora P, Kumar N, Puri A. Development of novel ionic liquid-based microemulsion formulation for dermal delivery of 5-fluorouracil. AAPS PharmSciTech. 2014;**15**(4):810-821. DOI: 10.1208/s12249-014-0103-1

[76] Agrawal OP, Agrawal S. An overview of new drug delivery system: Microemulsion. Asian Journal of Pharmaceutical Science and Technology. 2012;**2**(1):5-12

[77] Sarkhejiya Naimish A, Nakum Mayur A, Patel Vipul P, Atara Samir A, Desai TR. Emerging trend of microemulsion in formulation and research. International Bulletin of Drug Research. 2000;**1**(1):54-83

[78] Honda M, Asai T, Oku N, Araki Y, Tanaka M, Ebihara N. Liposomes and nanotechnology in drug development: Focus on ocular targets. International Journal of Nanomedicine. 2013;**8**:495. DOI: 10.2147%2FIJN.S30725

[79] Chaudhary A, Barman A, Gaur PK, Mishra R, Singh M. A review on microemulsion a promising optimising technique used as a novel drug delivery system. International Research Journal of Pharmacy. 2018;**9**:47-52

[80] Sapra B, Thatai P, Bhandari S, Sood J, Jindal M, Tiwary A. A critical appraisal of microemulsions for drug delivery: Part I. Therapeutic Delivery. 2013;**4**(12):1547-1564. DOI: 10.4155/tde.13.116

[81] Sapra B, Thatai P, Bhandari S, Sood J, Jindal M, Tiwary AK. A critical appraisal of microemulsions for drug delivery: Part II. Therapeutic Delivery. 2014;**5**(1):83-94. DOI: 10.4155/tde.13.125

Perspective Chapter: Overview of Bio-Based Surfactant – Recent Development, Industrial Challenge, and Future Outlook

Nur Liyana Ismail, Sara Shahruddin and Jofry Othman

Abstract

Bio-based surfactants are surface-active compounds derived from oil and fats through the production of oleochemicals or from sugar. Various applications of bio-based surfactants include household detergents, personal care, agricultural chemicals, oilfield chemicals, industrial and institutional cleaning, and others. Due to the stringent environmental regulations imposed by governments around the world on the use of chemicals in detergents, as well as growing consumer awareness of environmental concerns, there has been a strong demand in the market for bio-based surfactants. Bio-based surfactants are recognized as a greener alternative to conventional petrochemical-based surfactants because of their biodegradability and low toxicity. As a result, more research is being done on producing novel biodegradable surfactants, either from renewable resources or through biological processes (bio-catalysis or fermentation). This chapter discusses the various types, feedstocks, and applications of bio-based surfactants, as well as the industrial state-of-the-art and market prospects for bio-based surfactant production. In addition, relevant technological challenges in this field are addressed, and a way forward is proposed.

Keywords: bio-based surfactant, green surfactant, biosurfactant, renewable materials, sustainable surfactant

1. Introduction

Surfactants are surface-active agents that reduce water–oil, liquid–gas, and solid–liquid or solid–gas medium surfaces and interfacial tension [1, 2]. The surface energy is reduced by the presence of hydrophilic and hydrophobic sections of the same surfactant molecule owing to preferred interactions at surfaces and interfaces. In aqueous solution, surfactant molecules arrange themselves at the interface, where the hydrophobic part is in the air (or oil) and the hydrophilic part is in water, while at high concentration or concentrations above the critical micelle concentration (CMC), surfactant molecules self-assemble into micelles (**Figure 1**). Not only are they widely used as cleaning agents, but also other beneficial properties, such as foaming, emulsification, and particle suspension, make surfactants known for their wetting ability and effectiveness such as emulsifiers and stabilizers. Due to this characteristic, surfactants are found in a variety of products that we use every day, including food, pharmaceuticals, toiletries, detergents, automotive fluids, paints,

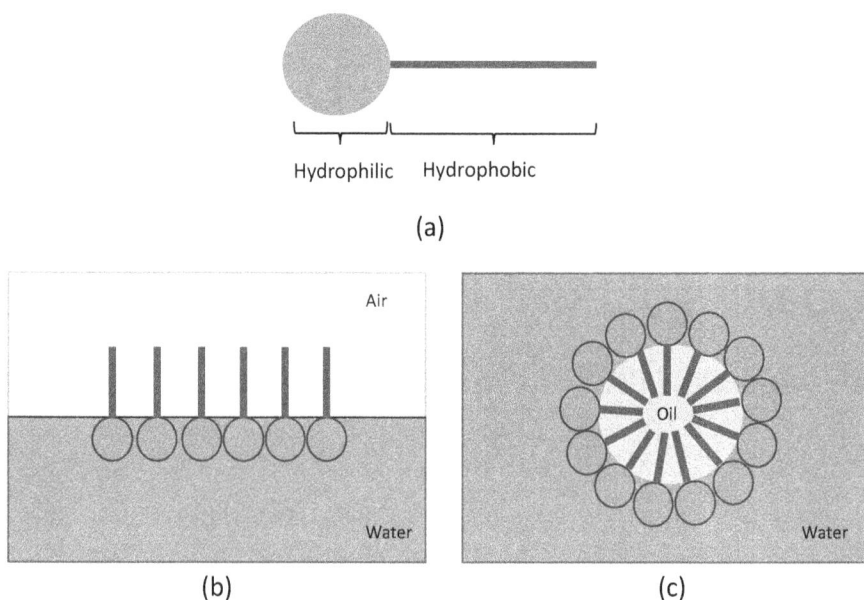

Figure 1.
(a) Simplified surfactant molecule, (b) arrangement of surfactant monomers at the water surface, and (c) micelle formation above critical micelle concentration (CMC).

and coatings [2]. Surfactants have steadily grown in popularity since their debut in the early twentieth century, and they are now among the most widely used synthetic compounds on the planet [3, 4].

Petrochemical and renewable sources are the two primary feedstock groups used in the manufacture of surfactants [5, 6]. The development of petrochemical processing led to the acquisition of hydrophobic structures of surfactant molecules through polymerization of alkenes, such as ethylene or propylene. Although ethylene has been employed as a carbon chain-building block, its increased applicability in industrial production has resulted from the production of an intermediate or precursor, ethylene oxide [7]. Natural surfactants are usually derived from triglycerides found in vegetable oils or animal fats. The surfactant industry was focused on the saponification of oils and fats prior to petrochemical processing [8, 9]. Surfactants infiltrate water bodies after usage, where they can create issues if they remain for a long time, resulting in the buildup of potentially toxic or otherwise hazardous substances causing significant environmental concerns [10–12]. Synthetic surfactant-related water contamination has increased in recent years because of its widespread usage in domestic, agricultural, and other cleaning activities. This occurrence has caused global concern, forcing establishment of a series of new rules governing its usage and disposal [13, 14]. In addition, experts relate the production of petrochemical-based surfactants to the high net output of CO_2, a greenhouse gas linked to climate change and global warming. By switching to renewable feedstock, this rate can be minimized. A previous study shows that using renewable resources instead of petrochemicals for surfactant synthesis would cut CO_2 emissions by 37% in the EU [15]. Beside environmental concerns and regulations, growing consumer awareness and market pressures have prompted considerable R&D into bio-based surfactants as potential substitutes for synthetic surfactants.

The term "bio-based surfactant" refers to a surfactant produced by a chemical or enzymatic process that uses renewable substrates as raw materials [16, 17]. According to ISO/DIS 21680, a bio-based surfactant is defined as a surfactant

Surfactant class	Bio-based carbon content X% (*m/m*)
Wholly bio-based surfactant	≥95
Majority bio-based surfactant	95 ≥ X > 50
Minority bio-based surfactant	50 ≥ X ≥ 5
Non-bio-based surfactant	X < 5

Table 1.
Bio-based surfactant classes according to CEN/TS 17035 [19].

wholly or partly derived from biomass (based on biogenic carbon) [18]. Most applications need further processing of bio-based feedstocks to incorporate functional groups that can give the surfactant's functional characteristics, resulting in a variety of anionic, cationic, nonionic, and amphoteric products. Many of these processes require the use of petroleum-based feedstocks or moieties that are not always environmentally friendly. The European Commission of Standardization has created categories for biosurfactants, including >95% completely bio-based, 50–94% majority bio-based, 5–49% minority bio-based, and 5% non-bio-based to assist in analyzing the bio-based surfactants' sustainability criteria (**Table 1**) [19].

The hydrophobe, hydrophile, or both, which are derived from natural sources, can be used in the production of bio-based surfactants. Plant oil, fatty acids, and animal fat are examples of natural hydrophobes, while glycerol, glucose, sucrose, and amino acids (aspartame, glutamic, lysine, arginine, alanine, and protein hydrolysates) are examples of natural hydrophiles. They can be either directly utilized in their original form or produced from complicated sources, such as vegetable oil, sugarcane, sugar beets, and starch-producing crops. As for biosurfactants, they consist of hydrophilic sugar or peptide component and hydrophobic saturated or unsaturated fatty acid chains that are naturally produced by bacteria, yeast, and fungi. Hence, a biosurfactant is classified as a wholly bio-based surfactant since all its raw materials are considered natural [20–22].

The hydrophobic part of bio-based and biosurfactant feedstock is mostly from fatty acyl groups. The fatty acyl groups are generally obtained from oilseeds in the form of triacylglycerol, but they may also be derived from oleochemical by-products such as free fatty acid or phospholipids. Fatty acyl groups are generally utilized as lipophilic building blocks for surfactants in the form of free fatty acids or fatty acyl esters, which are produced *via* hydrolysis or alcoholysis of triacylglycerol [23, 24]. This fatty acyl group conjugates hydrophilic and lipophilic compounds *via* an ester bond. This bond makes the fatty acid-based surfactants suitable for foods, cosmetics, personal care, and pharmaceutical product applications, but not for laundry detergents since the ester bonds are unstable. More stable bonds, such as ether, amides, and carbonate bonds, can be produced by converting the fatty acid groups to fatty alcohols, fatty amines, or fatty acid chloride [25–27].

Algae are another potential renewable source of fatty acids. It has been an active research area in recent years due to its potential for high oil production per acre and the ability to leverage on nonarable soil [28–30]. Previously, Unilever has partnered with Solazyme, a microalgae firm, with the aim of finding a palm-oil-free replacement for its soaps and surfactants. Solazyme used the advantage of its intellectual property in the areas of recombinant DNA expression in algae and algae bioprocessing to create oils with specific fatty acyl compositions [31]. Solazyme, later renamed as TerraVia, was acquired by Carbion in 2017 to focus on delivering innovative and high-value ingredients for food, personal care, and industrial applications [32]. Lignin has also been used as a feedstock in surfactant production due to its hydrophobic aromatic structure. Lignin-based surfactants are usually made by grafting hydrophilic groups or monomers onto the lignin to enhance its surface properties [33–35]. Extensive investigations are necessary to expedite the commercialization of lignin-based surfactants to the market since information on connecting performance and characteristics of lignin-based surfactants for their optimal usage is still lacking.

Among the most significant feedstocks for renewable hydrophile sources are vegetable oils (for glycerol), sugarcane and sugar beets (for sucrose), and starch-producing crops, such as maize, wheat, potato, and tapioca (for glucose) [4, 23, 36]. The use of glycerol as an alternative hydrophilic building block to replace ethylene oxide in the synthesis of nonionic surfactants is a feasible option. The major glycerol-based surfactants in the market are ester-based mono- and diglycerides, which are made by transesterifying triglycerides with excess glycerol and a base catalyst [4, 26, 37]. Carbohydrates, such as sugar and sucrose, are another useful biorefinery feedstock that make up as surfactant hydrophiles. The discovery of sucrose monoesters, or long-chain fatty acid esters, was one of the first major achievements of the Sugar Research Foundation (SRF) and led to their use as nonionic surfactants, food additives, and emulsifiers [38]. The global sucrose esters market amounted to $71.9 M in 2018 and is expected to reach $137.85 M by 2027 [39]. However, selectivity in the synthesis of these esters remains a challenge where acylation with a single fatty acid can yield many different isomers with various degrees of substitution [40]. One of the solutions to tackle the selectivity problem is by using lipases and proteases for regioselective sucrose ester production [41, 42]. Further improvement *via* lipase and protease protein engineering might increase the regioselectivity and yield of the catalysis processes. The biotransformation of sucrose to sucrose esters utilizing whole-cell fermentation methods might also give a new path to sucrose-based surfactant production.

Glucose is utilized as a hydrophile in the manufacture of a variety of surfactants, both directly and indirectly. It can react directly with fatty alcohol in a glycosidation process to produce alkyl polyglucosides (APGs), a nonionic surfactant class with growing production and popularity. Indirectly, glucose may be chemically converted to sorbitol, sorbitan, N-methyl glucamine, and O-methylglucoside, or enzymatically converted to amino, lactic, and citric acids, all of which can be leveraged to produce surfactants (**Figure 2**) [4].

Sugar-derived surfactants have a higher market demand than synthetic chemicals and surfactants due to their low toxicity, low cost, biodegradability, good cleaning and washing abilities, environmental compatibility, and high surface activity [43, 44]. However, if the demand for sugar surfactants grows in the long run, feedstock availability will become a concern. New methods that use bacteria and microorganisms to manufacture glucose are emerging; however, the issue of scalability has yet to be solved.

The creation of new amino acid-based surfactants may be influenced by advancements in biotechnological amino acid synthesis. Other than L-glutamic acid and L-lysine, which are the two most produced amino acids in the market, alanine, aspartic acid, glycine, and arginine, as well as protein hydrolysates, are also used in the manufacture of some commercial surfactants [45–47]. Another type of amino acid surfactant, sarcosine-based surfactants, has been in the market for decades. Even though sarcosine is a naturally occurring molecule, it is mostly synthesized on a large scale by combining chloroacetic acid with N-methylamine [48–50]. Betaine, another naturally occurring molecule, is also synthesized in large scale using petrochemical-based trimethylamine and chloroacetic acid. Most betaine surfactants use an oleochemical hydrophobe precursor obtained from tropical oils as the bio-based component [51]. Glycine betaine is a promising biosurfactant that can be commercially extracted from brown algae and sugar beet molasses [52, 53].

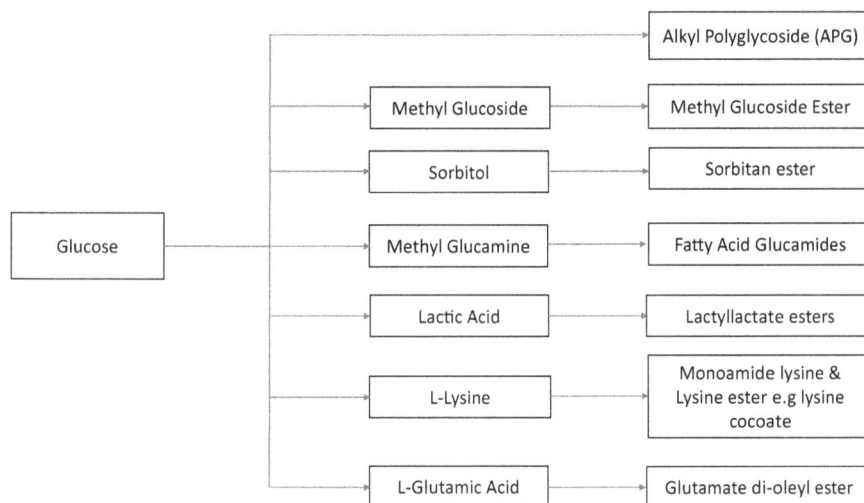

Figure 2.
Simplified transformations pathway from glucose to several surfactant building blocks and surfactants.

Glycolipids are a type of complex carbohydrate that contains both a glycan and a lipid component. They are usually the main lipids of bacterial and fungal cell walls. In an aqueous solution, glycolipids are amphiphilic substances that form stable micelles, and these molecules have the capacity to offer low interfacial tension [54, 55]. Rhamnolipids and sophorolipids are among the glycolipids that have been utilized the most as biosurfactants. Rhamnolipids are produced as one or two rhamnose sugar groups attached to one or two fatty acid chains by different bacterial species (i.e., *Pseudomonas aeruginosa*, *Pseudomonas chlororaphis*, *Burkholderia pseudomallei*) [4, 56]. Beside their favorable emulsifying, solubilizing, foaming, and antibacterial characteristics, the use of rhamnolipids is appealing due to their high production yields after relatively short incubation times [56]. Rhamnolipids are now available on a larger scale due to the optimized fermentation techniques and advanced extraction and concentration technologies. Sophorolipids, another extensively researched type of glycolipid, are biosynthesized by certain yeast strains such as *Starmerella bombicola*, *Wickerhamiella domercqiae*, and *Candida batistae* from sophorose sugar and hydroxylated fatty acid. Sophorolipids are commercially used in dish and vegetable detergents and in skin care formulations [57–60].

2. Recent progress in R&D and industrial production

Regulations on the environmental impact and hazardous chemicals are highly stringent, particularly in Europe and North America, which are the two largest markets for surfactants, especially in the home and personal care sectors. As a result, the surfactant industry is commencing to develop biosurfactants, which have lower levels of toxicity and a more environmentally friendly manufacturing process. Apart from complying with environmental regulations, the industry is seeing bio-based surfactants to achieve a sustainable competitive edge. The advent of biotechnology in the twenty-first century promoted the creation of novel bio-based and biosurfactants along with their enhanced commercial and economic viability. Extensive and significant R&D has also enabled high-quality and high-functionality bio-based surfactant formulations to evolve from the lab scale to niche applications to commercial-scale production. Some of the bio-based surfactants that are commercially available in the market, their main manufacturers, and their applications are listed in **Table 2**.

In the current development of novel surfactants, there is a growing trend of utilizing nontraditional naturally occurring branching hydrophobic chains [61–63]. Nonionic surfactants based on twin tail glycerol have been synthesized and they have good oil-in-water and water-in-oil emulsifying characteristics [64]. Other

Bio-based surfactants	Selected manufacturers	Fields of applications
Anionic		
Lignosulfonate Methyl ester sulfonates Anionic derivatives of alkyl polyglucoside	Vanderbilt Minerals, LLC Huish Detergents, Inc., Lion Corp., Longkey, Stepan Cognis, Colonial Chemical	• Laundry • Food service and kitchen hygiene • Dishwashing • Hard surface cleaning • Institutional cleaning and sanitation • Vehicle and transportation care
Nonionic		
Fatty alcohol alkoxylate Fatty acid alkoxylate Alkyl polyglucoside Sorbitan ester Alkanoyl-*N*-methylglucamide Alkyl ethoxylated mono- and diglycerides Polyglycerol esters	BASF, Dow BASF, Clariant, Croda, Croda, Esterchem, Huntsman Akzo Nobel, BASF, Colonial Chemical, Dow, Huntsman BASF, Croda, Huntsman Clariant, Kao, Kerry Ingredients and Flavors BASF, Colonial Chemical, Hychem Corp., Kerry Ingredients and Flavors	• Dishwashing • Laundry • Hard surface cleaning • Food service and kitchen hygiene • Institutional cleaning and sanitation • Vehicle care • Personal care
Amphoteric		
Cocoamidopropyl betaine Cocoamidopropyl hydroxysultaine Lauryl hydroxysultaine	BASF, Colonial Chemical, Stepan Colonial Chemical, Stepan Colonial Chemical, Stepan	• Hard surface cleaning • Food and beverage processing • Personal care
Glycolipid		
Sophorolipid Rhamnolipid	BASF, Clariant, Ecover, Evonik, MG Intobio Co. Ltd., Saraya Co. Ltd., Soliance AGAE Technologies, BASF, Biotensidon GmbH, Clariant, Evonik, GlycoSurf, Henkel, Jeneil Biotech Inc., Logos Technology Rhamnolipid Companies Inc., TeeGene Biotech	• Personal care • Vegetable liquid wash • Dish washing
Amino acid surfactants		
Sodium cocoyl glutamate Sodium methyl cocoyl taurate α-Acyl glutamate and sarcosinate	Ajinomoto Co. Inc., Stepan, Zschimmer and Schwarz Clariant Schill+Seilacher	• Personal care

Table 2.
Commercially available bio-based surfactants, their manufacturers, and their applications.

structural analogs of glycerol-based surfactants have recently been created by employing heterogeneous interfacial acidic catalysts to directly etherify glycerol and dodecanol. These surfactants have been shown to be comparable with commercially available surfactants in terms of physicochemical assessment and detergency ability [37]. Another class of amphiphilic compounds with a glycerol backbone is bio-based dialkyl glycerol ethers. These compounds have good solvo-surfactant characteristics and can function as solubilizers for hydrophobic dyes in aqueous media [65].

Natural edible flavor vanillin is used to create a cleavable vanillin-based polyoxy-ethylene nonionic surfactant. Because it contains cleavable acetal bonds that break down quickly under acidic circumstances, this environmentally beneficial surfactant is totally biodegradable in nature. The surfactant's surface activity, wettability, and emulsifying and foaming properties are on par with nonylphenol ethoxylate surfactants, which are highly toxic to aquatic organisms and environment [66]. Several novel types of sustainable surfactant have been created in recent years by employing various types of terpenes, which are the major components of essential oils derived from a variety of plants and flowers [67–70]. The terpenes were transformed to branched hydrophobic tail containing quaternary ammonium surfactants. Natural farnesol, a 15-carbon acyclic sesquiterpene alcohol found in neroli, lemongrass, tuberose, rose, citronella, and other plant species, was used to create a new form of terpene-based sustainable surfactant, which has demonstrated excellent surfactant performance [70]. Under the trade name ECOSURF, Dow Chemical Co. is now offering a range of sustainable oilseed-based nonionic surfactants. These surfactants are claimed to have minimal aquatic toxicity and are biodegradable in nature, making them suitable candidates for paints and coatings, as well as home, industrial, and institutional cleansers and textiles [71].

TegraSurf, a range of sustainable water-based surfactants developed for energy, mining, agricultural, water treatment, and other industrial applications, was released in July 2021 by Integrity BioChem (IBC), a technology-driven business producing next-generation biopolymers. TegraSurf is made of sustainable vegetal materials and is certified Readily Biodegradable by the OECD 301B guideline. After 90 days, it is no longer present in the environment, making it safer and healthier for local populations and allowing formulators to fulfill industry sustainability criteria [72]. BASF and Solazyme Inc. recently released Dehyton® AO 45, the first commercial microalgae-derived betaine surfactant made from microalgae oils as an alternative to conventional amidopropyl betaine surfactants [73]. Following the launch of sophorolipid-based surfactants in 2020, BASF formed an exclusive partnership with Holiferm Ltd. in the United Kingdom to focus on the development of glycolipids other than sophorolipids for personal and home care as well as for industrial uses [74].

Croda expanded its commercial-scale bio-based manufacturing capabilities and technology leadership in renewable raw materials by unveiling its 100% bio-based ethylene oxide production facility as an effort to make the world's products greener. Ethylene oxide is the key raw material used to produce surfactants. Croda's Atlas Point manufacturing plant in New Castle, Delaware, is the first of its type in the United States for the manufacture of 100% sustainable, 100% bio-based nonionic surfactants [75]. Ajinomoto is increasing to 60% of its global capacity for its Amisoft range of amino acid-based liquid surfactants by building a new plant in Brazil [76]. Sironix Renewables received $645,000 in investment from the University of Minnesota Discovery Capital Investment program and investors as well as a $1.15 million grant from the US Department of Energy Advanced Manufacturing Office, to help them scale up their Eosix® production. The new renewable oleo-furans-based surfactants are 100% plant-based that offer unique and adjustable characteristics in a wide range of areas, including cleaning products, cleaners, cosmetics and personal care, agriculture and inks, and paint and coatings [77].

3. Industrial challenge on bio-based surfactant

This section covers the market performance, demand drivers, and growth prospects of biosurfactants. The market trend on bio-based and biosurfactants is discussed for the different geographic regions and in terms of changing market trends for biosurfactants in various application areas. Analysis of the industrial challenges of biosurfactants, which include the growth-restraining factors and future opportunities, is provided.

3.1 The economy and market trend of bio-based surfactant

The worldwide surfactant industry, estimated to be worth $39 billion in 2019, is expected to expand at a rate of 2.6% per year over the following five years, reaching $46 billion in 2024. Surfactants are produced in total of 17 million metric tons per year [78]. In the EU, of the 3 million metric tons of surfactants produced in 2019, roughly 50% were bio-based [79]. A market study by Market Research Future [80] indicated that the global biosurfactants' market value is around USD 2.1 billion in 2020 and predicted it to reach USD 2.8 billion by 2026, with a compound annual growth rate of over 5% from 2021 to 2026. The attractive performance of biosurfactants advances their high potential to substitute synthetic-based surfactants for drop-in applications and with unique properties that can overcome entry barriers for the emerging industrial areas. Major types of biosurfactants, such as sophorolipids, glycolipids, lipopeptides, polymeric biosurfactants, phospholipids and fatty acids, generally form the product demand application. Among biosurfactants, sophorolipids provide the largest global market demand with detergents and industrial cleaning applications. The leading demand drivers for biosurfactants comprise a growing consumer preference, increasingly stringent regulatory requirements, and rising awareness toward eco-friendly alternatives. By being environmentally compatible and with low toxicity, many studies have considered biosurfactants as the next generation of industrial surfactants [81–83]. In terms of end-user applications, biosurfactants are finding usage in household detergents, industrial and institutional cleaners, cosmetics, and personal care within the major markets in Europe and North America [80]. Recently, they have been gaining acceptance in the newer application areas such as in oil and gas as well as in agricultural industries.

Furthermore, the increasing consumer awareness of the benefits of biosurfactants and their wide range of application sectors form market drivers that increase their future growth potential. Higher growth of biosurfactants is seen in Asia-Pacific (APAC), especially in Southeast Asian countries that have slightly different demand factors that involve the increasing purchasing power of mass consumers, growing concern on environmental issues, and the generation of harmful chemical by-products. In terms of APAC market segmentation, the major sales revenue for biosurfactants resides within the home care and personal care applications, as rising urbanization becomes the dominant factor for surfactant growth. More importantly, a key growth enabler is in the innovative research on biosurfactants, especially when it can generate multifunctional and diversified products using renewable feedstock. This technological progress contributes to the desirable properties of biosurfactants to meet the changing consumer lifestyles in developing economies and consequently their increasing preference for usage in the end-user product formulation. As an example, within the home care detergent industry, the usage of biosurfactants as environmentally friendly products provides sustainable alternatives that are gaining a large market share [81, 84, 85].

The highest adoption of bio-based and biosurfactants is in Europe and North America, which dominate bio-based surfactant market share in terms of revenue and volume. Increasingly stringent regulatory requirements enable a wider acceptance of biosurfactants in the place of synthetic surfactants. For example, the imposed government regulations, such as CEN/TC-276, define the standards for surface-active agents and detergents to enhance the EU bio-based economy, detergent regulation (EC) No 648 that require surfactants used in detergents to be biodegradable under aerobic conditions as per OECD 301 test series. In addition, the COVID-19 pandemic results in a sharp increase in the bio-based surfactant product demand for household detergents, personal care, and industrial cleaners due to the rising trend for sanitation.

3.2 The industrial challenges of bio-based surfactant

Bio-based surfactants are synthesized *via* a chemical reaction, which is usually carried out under harsh conditions. The use of hazardous solvents and toxic acid or base catalysts sometimes creates undesired waste or by-products that are detrimental to the environment. Enzymes have the potential to play a significant role in the production of numerous bio-based surfactants, although they are not currently used on a large basis. Enzymes provide several advantages over chemical processing, notably in terms of improving process sustainability. The main drawbacks of enzymes are their relatively higher price compared to chemical catalysts as well as their slower reaction speeds. However, since energy costs are expected to rise, the need of sustainability (lower operating energy, less waste, and safer operating condition) is crucial. Despite the growing demand for bio-based surfactants, several challenges exist that restrain their further market growth and wider adoption. The main challenge is in the higher pricing of bio-based and biosurfactants as the biggest hurdle in meeting the requirement of priced sensitive Asian customers. Higher complexity and low-efficiency microbial fermentation process in biosurfactant manufacturing contribute to the high production cost and expensive capital cost investment. For example, the average price of sophorolipids is USD 34 per kilogram as compared to sodium dodecyl sulfate and amino acid surfactants that are priced at USD 1–4 per kilogram [86]. Nevertheless, a lower operating cost of USD 2530/ton for sophorolipids' production is attainable through technological improvement such as integrated separation, which places sophorolipid surfactants at similar prices to other specialty surfactants [87]. Increased sustainability of biosurfactant alone without significantly higher performance is not well accepted, as the usual consumers will not be willing to pay a "green" premium for bio-based products. Therefore, lower cost improvement in biosurfactant manufacturing is fundamentally important to attain an economically sustainable process and assure future market continuity [85].

A second challenge is the dependency of biosurfactant demand on the volatility and economic downturn of downstream end user industries. Industries that are applicable for biosurfactant applications, such as oil and gas, enhanced oil recovery, food industry, construction, textiles, paints, pharmaceutical, and detergents, are known to be susceptible to general macroeconomic performance. In addition, the COVID-19 pandemic further leads to disruption in the end-user industrial demand and sustainability concern on the raw material supply. The sustainability of raw materials is a major concern as these contribute up to 50% of the glycolipid production cost and 10–30% of the overall cost for other biosurfactant products. Purification accounts for 60% of the production cost, but this can be minimized for the case of biosurfactant application in crude forms, such as in an industrial environment [88]. However, for high-purity applications, improvement in downstream processing methods is needed to attain a competitive cost of production. Opportunity exists in developing a new technology solution that utilizes a low-cost raw material such as industrial wastes for biosurfactant production. However, this needs to consider the overall production impact factors that include the availability, stability, and variability of each component [88]. The economic viability criteria for biosurfactant production, therefore, include microorganism performance, bioreactor design, target market, purification process, product properties, production condition, fermentation cycle time, and production yield [89].

Additionally, several operation and control factors provide important handles to minimize biosurfactant production costs. Batch cycle optimization on the fermentation and purification process can reduce the idle time between batches and minimize chemical usage for equipment cleaning and energy use during sterilization. Productivity is the most important factor in the manufacturing economics of biosurfactant production at commercial scales [8]. Optimum batch-sequencing campaign minimizes startup and shutdown frequency to lower the production downtime that improves productivity. Lastly, biosurfactant product development will need to fulfill time-consuming and expensive legislative requirements, which restrain market growth [90]. These add a high cost of compliance to the product development cost that is incurred by biosurfactant manufacturers. Other market entry requirements include the biosurfactant products that are tested for long shelf life and the ability to maintain stable properties in the industrial environment [91].

4. Future outlook and prospect

The development of bio-based surfactants from renewable feedstocks is an attractive alternative to fossil-based surfactants with a significantly growing market attributed to their performance, biodegradability, biocompatibility, and nontoxicity [22, 33]. Additionally, advances in renewable technology, increased environmental concern, consumer awareness, and stringent regulatory requirements provide a continued push toward the demand of bio-based surfactants. Potential areas for use are growing fast, and valuable outcomes depend on whether the bio-based surfactants can be customized for specific applications along with if they can be produced at a price that will make them attractive alternatives to the fossil-based surfactants. The simultaneous design of bio-based surfactants for functional, economic, and environmental benefits will be taxing, but it will ensure the replacement of conventional fossil-based surfactants provided they can offer comparable or superior performance and a unique value proposition.

Presently, fossil-based surfactants are less expensive than bio-based surfactants [4, 92, 93]. However, this trend will likely change in the future, thereby increasing the prospects of bio-based surfactants. Feedstocks and how the bio-based surfactants are produced are the two key factors governing final product costs [4, 36, 94, 95]. To use renewable feedstock in the industry, they should be cost-effective, available in large quantities, and can effectively be converted to value-added surfactants [95]. Renewable feedstocks used as starting materials to produce surfactants usually face severe economic competition from their fossil-based counterparts. Surfactants comprised of hydrophilic head group and hydrophobic tail group, which are linked by a chemical bond generating an amphiphilic molecule that can be used directly or further modified. Surfactant design requires careful selection of the hydrophile and hydrophobe pair so that they can be easily synthesized with minimum purification and provide the desired properties for the intended application [4, 16, 92, 96]. Triglycerides, fatty acid methyl esters, fatty alcohols, fatty acids, and fatty amines are common examples of renewable hydrophobes used to produce bio-based surfactants. Sustainable hydrophilic headgroups can be designed using several molecules such as glycerol, carbohydrate feedstocks such as sucrose, glucose, organic acids, and amino acids [4, 36, 94, 95]. Additionally, the use of renewable feedstock for surfactant manufacturing also helps reduce CO_2 emissions because once the bio-based surfactants degrade, they only release back the quantitative amount of the carbon used by the plant to produce the surfactants [36]. Other than the starting material mentioned above, the use of alternative substrates, such as agro-based industrial wastes or other suitable simple waste substrate, is gaining a lot of research interest and can lead to significant cost reduction [97].

Researchers are continually improving the cost-effectiveness of production methods as well as enhancing the current technologies with green manufacturing principles to convert renewable feedstocks into valuable and new biobased surfactants. Some of the key focus areas include developing biobased surfactants from cheaper feedstocks, higher performance catalysts, green solvents, optimized reaction processes, and effective downstream purification could entice the industry players and end-use customers to make the switch from fossil-based surfactants to biobased surfactants. Catalyst design is also crucial to ensure high selectivity of the processes to limit or eliminate the formation of by-products and to help push the reaction forward towards completion faster [98–100]. Other than that, researchers are looking into equipment miniaturization such as continuous reactors to help reduce the raw material consumption and effluent production. Process intensification is another aspect that could help to reduce the investment costs [99]. Research focusing on alternative or green solvents dedicated to the conversion of renewable feedstock to value-added products has led to several publications. Among those being researched include bio-based ionic liquids, deep eutectic solvents, bio-based solvents, CO_2-switchable solvents and supercritical fluids [101–103].

In terms of market penetration of bio-based surfactants, customers tend to choose cost-effective surfactants. Despite much progress in technical knowledge, the large-scale production of bio-based surfactants using the methods described above is still limited. The commercial production of bio-based surfactants still faces many challenges that must be addressed for them to be economically viable. One major obstacle is the homogeneity and consistency of the feedstock, which can lead to inconsistency in the final bio-based surfactants. Variation in the surfactant properties and performance could lead to unsatisfactory properties. Thorough testing on the use of bio-based surfactants in place of fossil-based ones will also be needed to provide enough and convincing data on the merits of bio-based surfactants. It is hoped that these efforts will lead to broader use of bio-based surfactants in the future, offering enormous benefits such as excellent physicochemical properties, biodegradability, lower risk to human health, and minimum harm to the environment.

5. Conclusions

Surfactant manufacturers have introduced numerous new eco-friendly surfactant-based products to the market in the past few years. Increased consumer awareness, along with a responsibility for sustainable development, has resulted in the creation of several novel surfactant types based on renewable building blocks. These surfactants have improved biodegradation characteristics and low toxicity, making them a preferred alternative for innovative formulations in the industrial and consumer markets. However, these "drop-in" surfactant molecules, which aim to directly replace their petrochemical-based equivalents, face a huge challenge since prices must be as competitive as their fossil counterparts. Moreover, while several personal care and consumer product businesses have shown interest in 100% bio-based surfactants, only a few green premium products have been accepted into the market. More assessments and surveys need to be done to gauge consumer willingness to pay premium prices for other than commodity products. With increasing innovative formulations to meet consumer, legislative, and sustainability demands, it is obvious that the global demand for both petroleum- and bio-based surfactants will continue to grow, while manufacturers are challenged to balance cost-effective formulations with efficient performance.

Author details

Nur Liyana Ismail*, Sara Shahruddin and Jofry Othman
PETRONAS Research Sdn. Bhd, Kajang, Selangor, Malaysia

*Address all correspondence to: nurliyana.ismail@petronas.com.my

IntechOpen

References

[1] Rosen MJ, Kunjappu JT. Surfactants and Interfacial Phenomena. 4th ed. Hoboken, New Jersey: John Wiley & Sons; 2012

[2] Möbius D, Miller R, Fainerman VB. Surfactants: Chemistry, Interfacial Properties, Applications (Studies in Interface Science, Vol. 13). Amsterdam: Elsevier Science

[3] Traverso-Soto JM, González-Mazo E, Lara-Martín PA. Analysis of surfactants in environmental samples by chromatographic techniques. In: Calderon LA, editor. Chromatography – The Most Versatile Method of Chemical Analysis. London: IntechOpen; 2012

[4] Foley P, Beach ES, Zimmerman JB. Derivation and synthesis of renewable surfactants. Chemical Society Reviews. 2012;**41**:1499-1518

[5] Jimoh AA, Lin J. Biosurfactant: A new frontier for greener technology and environmental sustainability. Ecotoxicology and Environmental Safety. 2019;**184**:109-607. DOI: 10.1016/j.ecoenv.2019.109607

[6] Taddese T, Anderson RL, Bray DJ, et al. Recent advances in particle-based simulation of surfactants. Current Opinion in Colloid & Interface Science. 2020;**48**:137-148. DOI: 10.1016/j.cocis.2020.04.001

[7] Solar Illia CJDEAA, Sanchez C. Interactions between poly(ethylene oxide)-based surfactants and transition metal alkoxides: Their role in the templated construction of mesostructured hybrid organic-inorganic composites. New Journal of Chemistry. 2000;**24**(7):493-499. DOI: 10.1039/b002518f

[8] Zoller U, Sosis P, editors. Handbook of Detergents, Part F: Production. Boca Raton: CRC Press; 2008

[9] Holmberg K. Natural surfactants. Current Opinion in Colloid & Interface Science. 2001;**6**:148-159

[10] Deschênes L, Lafrance P, Villeneuve JP, Samson R. Adding sodium dodecyl sulfate and *Pseudomonas aeruginosa* UG2 biosurfactants inhibits polycyclic aromatic hydrocarbon biodegradation in a weathered creosote-contaminated soil. Applied Microbiology and Biotechnology. 1996;**46**(5):638-646

[11] Takeda K, Sasaoka H, Sasa K, Hirai H, Hachiya K, Moriyama Y. Size and mobility of sodium dodecyl sulfate—Bovine serum albumin complex as studied by dynamic light scattering and electrophoretic light scattering. Journal of Colloid and Interface Science. 1992;**154**(2):385-392

[12] Rebello S, Asok AK, Mundayoor S, Jisha MS. Surfactants: Toxicity, remediation and green surfactants. Environmental Chemistry Letters. 2014;**12**(2):275-287

[13] Siwayanan P, Bakar NA, Aziz R, Chelliapan S, Siwayanan P. Exploring Malaysian household consumers acceptance towards eco-friendly laundry detergent powders. Asian Social Science. 2015;**11**(9):125-137

[14] Farn RJ, editor. Chemistry and Technology of Surfactants. Oxford: John Wiley & Sons; 2008

[15] Patel M. Surfactants based on renewable raw materials: Carbon dioxide reduction potential and policies and measures for the European Union. Journal of Industrial Ecology. 2003;**7**(3-4):47-62

[16] Hayes DG, Smith GA. Bio-based surfactants: Overview and industrial state of the art. In: Hayes DG, Solaiman DK, Ashby RD, editors. London: Bio-Based Surfactants; 2019. pp. 3-38

[17] Kandasamy R, Rajasekaran M, Venkatesan SK, Uddin M. New trends in the biomanufacturing of green surfactants: Bio-based surfactants and biosurfactants. In: Rathinam NK, Sani RK, editors. Next Generation Biomanufacturing Technologies. Washington: American Chemical Society; 2019. pp. 243-260

[18] ISO/DIS 21680(en) Surface Active Agents—Bio-Based Surfactants—Requirements and Test Methods [Internet]. 2020. Available from: https://www.iso.org/obp/ui/#iso:std:iso:21680:dis:ed-1:v1:en

[19] Comite Europeen de Normalisation. CEN/TS 17035: 2017. Surface Active Agents—Bio-Based Surfactants—Requirements and Test Methods. Brussels: CEN. p. 2017

[20] Md F. Biosurfactant: Production and application. Journal of Petroleum & Environmental Biotechnology. 2012;3(4):124

[21] Van Renterghem L, Roelants SL, Baccile N, Uyttersprot K, Taelman MC, Everaert B, et al. From lab to market: An integrated bioprocess design approach for new-to-nature biosurfactants produced by *Starmerella bombicola*. Biotechnology and Bioengineering. 2018;115(5):1195-1206

[22] Moldes AB, Rodríguez-López L, Rincón-Fontán M, López-Prieto A, Vecino X, Cruz JM. Synthetic and bio-derived surfactants versus microbial biosurfactants in the cosmetic industry: An overview. International Journal of Molecular Sciences. 2021;22(5):2371

[23] Hayes DG, Solaiman DK, Ashby RD. Bio-Based Surfactants: Synthesis, Properties, and Applications. London: Elsevier; 2019

[24] Salimon J, Salih N, Yousif E. Industrial development and applications of plant oils and their bio-based

oleochemicals. Arabian Journal of Chemistry. 2012;5(2):135-145

[25] Kreutzer UR. Manufacture of fatty alcohols based on natural fats and oils. Journal of the American Oil Chemists' Society. 1984;61(2):343-348

[26] Biermann U, Bornscheuer U, Meier MA, Metzger JO, Schäfer HJ. Oils and fats as renewable raw materials in chemistry. Angewandte Chemie International Edition. 2011;50(17):3854-3871

[27] Giraldo L, Camargo G, Tirano J, Moreno-Pirajan JC. Synthesis of fatty alcohols from oil palm using a catalyst of Ni-Cu supported onto zeolite. E-Journal of Chemistry. 2010;7(4):1138-1147

[28] Jeon JY, Han Y, Kim YW, Lee YW, Hong S, Hwang IT. Feasibility of unsaturated fatty acid feedstocks as green alternatives in bio-oil refinery. Biofuels, Bioproducts and Biorefining. 2019;13(3):690-722

[29] Hess SK, Lepetit B, Kroth PG, Mecking S. Production of chemicals from microalgae lipids—Status and perspectives. European Journal of Lipid Science and Technology. 2018;120(1):1700152

[30] De Luca M, Pappalardo I, Limongi AR, Viviano E, Radice RP, Todisco S, et al. Lipids from microalgae for cosmetic applications. Cosmetics. 2021;8(2):52

[31] Lerayer A. Solazyme: "unlocking the power of microalgae: A new source of sustainable and renewable oils". In: BMC Proceedings. Vol. 8. BioMed Central; 2014. pp. 1-2

[32] Innovative Microalgae Specialist TerraVia Acquired by Corbion [Internet]. 2017. Available from: https://www.corbion.com/media/624357/20170929-tvia-completion-eng-final.pdf

[33] Alwadani N, Fatehi P. Synthetic and lignin-based surfactants: Challenges and opportunities. Carbon Resources Conversion. 2018;1(2):126-138

[34] Schmidt BV, Molinari V, Esposito D, Tauer K, Antonietti M. Lignin-based polymeric surfactants for emulsion polymerization. Polymer. 2017;112:418-426

[35] Zhou M, Wang W, Yang D, Qiu X. Preparation of a new lignin-based anionic/cationic surfactant and its solution behaviour. RSC Advances. 2015;5(4):2441-2448

[36] Bhadani A, Kafle A, Ogura T, Akamatsu M, Sakai K, Sakai H, et al. Current perspective of sustainable surfactants based on renewable building blocks. Current Opinion in Colloid & Interface Science. 2020;45:124-135

[37] Fan Z, Zhao Y, Preda F, Clacens JM, Shi H, Wang L, et al. Preparation of bio-based surfactants from glycerol and dodecanol by direct etherification. Green Chemistry. 2015;17(2):882-892

[38] Plat T, Linhardt RJ. Syntheses and applications of sucrose-based esters. Journal of Surfactants and Detergents. 2001;4(4):415-421

[39] Outlook for Sucrose Esters. Focus on Surfactants 2020;2020(6):3

[40] Queneau Y, Fitremann J, Trombotto S. The chemistry of unprotected sucrose: The selectivity issue. Comptes Rendus Chimie. 2004;7(2):177-188

[41] Potier P, Bouchu A, Descotes G, Queneau Y. Lipase-catalysed selective synthesis of sucrose mixed diesters. Synthesis. 2001;2001(03):0458-0462

[42] Cruces MA, Plou FJ, Ferrer M, Bernabé M, Ballesteros A. Improved synthesis of sucrose fatty acid monoesters. Journal of the American Oil Chemists' Society. 2001;78(5):541-546

[43] Jesus CF, Alves AA, Fiuza SM, Murtinho D, Antunes FE. Mini-review: Synthetic methods for the production of cationic sugar-based surfactants. Journal of Molecular Liquids. 2021;342:117389

[44] Hill K, Rhode O. Sugar-based surfactants for consumer products and technical applications. Lipid/Fett. 1999;101(1):25-33

[45] Infante MR, Pérez L, Pinazo A, Clapés P, Morán MC, Angelet M, et al. Amino acid-based surfactants. Comptes Rendus Chimie. 2004;7(6-7):583-592

[46] Takehara M. Properties and applications of amino acid based surfactants. Colloids and Surfaces. 1989;38(1):149-167

[47] Bordes R, Holmberg K. Amino acid-based surfactants—Do they deserve more attention? Advances in Colloid and Interface Science. 2015;222:79-91

[48] Baumann L. The preparation of sarcosine. Journal of Biological Chemistry. 1915;21(3):563-566

[49] Cullum DC. The separation of sarcosine from methylaminediacetic acid. Analyst. 1957;82(977):589-591

[50] Morris ED, Bost JC. Kirk-Othmer Encyclopedia of Chemical Technology. New Jersey: John Wiley & Sons, Inc.; 2002

[51] Clendennen SK, Boaz NW. Betaine amphoteric surfactants—Synthesis, properties, and applications. In: Hayes DG, Solaiman DK, Ashby RD, editors. Bio-Based Surfactants. London: AOCS Press; 2019. pp. 447-469

[52] Goursaud F, Berchel M, Guilbot J, Legros N, Lemiègre L, Marcilloux J,

et al. Glycine betaine as a renewable raw material to "greener" new cationic surfactants. Green Chemistry. 2008;**10**(3):310-320

[53] Nsimba ZF, Paquot M, Mvumbi LG, Deleu M. Glycine betaine surfactant derivatives: Synthesis methods and potentialities of use. Biotechnologie, Agronomie, Société et Environnement. 2010;**14**(4):737-748

[54] Bednarski W, Adamczak M, Tomasik J, Płaszczyk M. Application of oil refinery waste in the biosynthesis of glycolipids by yeast. Bioresource Technology. 2004;**95**(1):15-18

[55] Kakehi K, Suzuki S. Analysis of glycans; polysaccharide functional properties. In: Boons GJ, Lee YC, Suzuki A, Taniguchi N, Voragen AGJ, editors. Amsterdam: Comprehensive Glycoscience; 2007

[56] Irfan-Maqsood M, Seddiq-Shams M. Rhamnolipids: Well-characterized glycolipids with potential broad applicability as biosurfactants. Industrial Biotechnology. 2014;**10**(4):285-291

[57] Van Bogaert IN, Zhang J, Soetaert W. Microbial synthesis of sophorolipids. Process Biochemistry. 2011;**46**(4):821-833

[58] Hirata Y, Ryu M, Oda Y, Igarashi K, Nagatsuka A, Furuta T, et al. Novel characteristics of sophorolipids, yeast glycolipid biosurfactants, as biodegradable low-foaming surfactants. Journal of Bioscience and Bioengineering. 2009;**108**(2):142-146

[59] Develter DW, Lauryssen LM. Properties and industrial applications of sophorolipids. European Journal of Lipid Science and Technology. 2010;**112**(6):628-638

[60] Roelants S, Solaiman DK, Ashby RD, Lodens S, Van Renterghem L,

Soetaert W. Production and applications of sophorolipids. In: Hayes DG, Solaiman DK, Ashby RD, editors. London: Bio-Based Surfactants; 2019. pp. 65-119

[61] Iskandar WF, Salim M, Hashim R, Zahid NI. Stability of cubic phase and curvature tuning in the lyotropic system of branched chain galactose-based glycolipid by amphiphilic additives. Colloids and Surfaces A: Physicochemical and Engineering Aspects. 2021;**623**:126697

[62] Kim JH, Oh YR, Hwang J, Kang J, Kim H, Jang YA, et al. Valorization of waste-cooking oil into sophorolipids and application of their methyl hydroxyl branched fatty acid derivatives to produce engineering bioplastics. Waste Management. 2021;**124**:195-202

[63] Elsoud MM. Classification and production of microbial surfactants. In: Inamuddin, Ahamed MI, Prasad R, editors. Microbial Biosurfactants. Singapore: Springer; 2021. pp. 65-89

[64] Zhang L, Zhang X, Zhang P, Zhang Z, Liu S, Han B. Efficient emulsifying properties of glycerol-based surfactant. Colloids and Surfaces A: Physicochemical and Engineering Aspects. 2018;**553**:225-229

[65] Lebeuf R, Illous E, Dussenne C, Molinier V, Silva ED, Lemaire M, et al. ACS Sustainable Chemistry & Engineering. 2016;**4**:4815-4823

[66] Ding F, Zhou X, Wu Z, Xing Z. Synthesis of a cleavable vanillin-based polyoxyethylene surfactant and its pilot application in cotton fabric pretreatment. ACS Sustainable Chemistry & Engineering. 2019;**7**:5494-5500

[67] Ogunkunle T, Fadairo A, Rasouli V, Ling K, Oladepo A, Chukwuma O, et al. Microbial-derived bio-surfactant using neem oil as substrate and its suitability for enhanced oil recovery. Journal of

Petroleum Exploration and Production Technology. 2021;**11**(2):627-638

[68] Ma J, Gao J, Wang H, Lyu B, Gao D. Dissymmetry gemini sulfosuccinate surfactant from vegetable oil: A kind of environmentally friendly fatliquoring agent in the leather industry. ACS Sustainable Chemistry & Engineering. 2017;**5**:10693-10701

[69] Faßbach TA, Gaide T, Terhorst M, Behr A, Vorholt AJ. Renewable surfactants through the hydroaminomethylation of terpenes. ChemCatChem. 2017;**9**:1359-1362

[70] Bhadani A, Rane J, Veresmortean C, Banerjee S, John G. Bio-inspired surfactants capable of generating plant volatiles. Soft Matter. 2015;**11**: 3076-3082

[71] ECOSURF™ SA Surfactants-Seed Oil-Based Surfactants [Internet]. Dow; 2008. Available from: https://www.dow.com/documents/en-us/mark-prod-info/119/119-02222-01-ecosurf-sa-surfactants-seed-oilbased-surfactants.pdf

[72] Integrity BioChem Announces First-of-its-Kind Surfactant to Improve Industrial Sustainability [Internet]. Businesswire; 2021. Available from: https://www.businesswire.com/news/home/20210728005149/en/Integrity-BioChem-Announces-First-of-its-Kind-Surfactant-to-Improve-Industrial-Sustainability

[73] BASF and Solazyme Launch the First Commercial Microalgae-Derived Betaine Surfactant [Internet]. BASF; 2015. Available from: https://www.basf.com/us/en/media/news-releases/2015/07/P-US-15-137.html

[74] How Bio-Based Surfactants are Turning the World Green [Internet]. Cosmetics & Toiletries; 2020. Available from: https://www.cosmeticsandtoiletries.com/formulating/category/natural/

How-Bio-Based-Surfactants-are-Turning-the-World-Green---570209271.html

[75] BASF Strengthens Its Position in Bio-Surfactants for Personal Care, Home Care and Industrial Formulators with Two Distinct Partnerships. BASF; 2021. Available from: https://www.basf.com/my/en/media/news-releases/global/2021/03/p-21-148.html

[76] Ajinomoto to Expand Amino Acid-Based surfactants [Internet]. Specialty Chemicals Magazine; 2019. Available from: https://www.specchemonline.com/index.php/ajinomoto-expand-amino-acid-based-surfactants

[77] Sironix Renewables Closes Seed Round to Scale Production of its Plant-Based Surfactant for Detergents, Cleaners & Shampoos [Internet]. GlobeNewswire; 2020. Available from: https://www.globenewswire.com/news-release/2020/09/16/2094579/0/en/Sironix-Renewables-Closes-Seed-Round-to-Scale-Production-of-its-Plant-based-Surfactant-for-Detergents-Cleaners-Shampoos.html

[78] Assessing the Sustainability and Performance of Green Surfactants [Internet]. IHS Markit; 2020. Available from: https://ihsmarkit.com/research-analysis/assessing-sustainability-and-performance-of-green-surfactants.html

[79] Bio-Based Chemicals—A 2020 Update. IEA Bioenergy Task 42 Report [Internet]. 2020. Available from: https://www.ieabioenergy.com/wp-content/uploads/2020/02/Bio-based-chemicals-a-2020-update-final-200213.pdf

[80] Market Research Future. 2021. Available from: https://www.reportlinker.com/p06081965/Biosurfactants-Market-Research-Report-by-Product-Type-by-Application-by-Region-Global-Forecast-to-Cumulative-Impact-of-COVID-19 [Accessed: 15 August 2021]

[81] Rocha e Silva NMP, Meira HM, FCA A, et al. Natural surfactants and their applications for heavy oil removal in industry. Separation and Purification Reviews. 2019;**48**(4):267-281. DOI: 10.1080/15422119.2018.1474477

[82] Drakontis CE, Amin S. Biosurfactants: Formulations, properties, and applications. Current Opinion in Colloid & Interface Science. 2020;**48**:77-90. DOI: 10.1016/j.cocis.2020.03.013

[83] Naughton PJ, Marchant R, Naughton V, et al. Microbial biosurfactants: Current trends and applications in agricultural and biomedical industries. Journal of Applied Microbiology. 2019;**127**(1): 12-28. DOI: 10.1111/jam.14243

[84] Almeeida DG, Sores da Silva RCF, Luna JM, et al. Biosurfactants: Promising molecules for petroleum biotechnology advances. Frontiers in Microbiology. 2016;**7**:1718

[85] Silva RCFS, Almeida DG, Luna JM, et al. Applications of biosurfactants in the petroleum industry and the remediation of oil spills. International Journal of Molecular Sciences. 2014;**15**(7):12523-12542. DOI: 10.3390/ijms150712523

[86] Rodriguez-Lopez L, Rincon-Fontan M, Vecino X, et al. Extraction, separation and characterization of lipopeptides and phospholipids from corn steep water. Separation and Purification Technology. 2020;**248**. DOI: 10.1016/j.seppur.2020.117076

[87] Roelants S, Van Renterghem L, Maes K, Evaraert B, Vanlerberghe B, Demaeseneire S, et al. Mirobial biosurfactants: From lab to market. In: Press C, editor. Microbial Biosurfactants and Their Environmental and Industrial Applications. Boca Raton: CRC Press; 2019. DOI: 10.1002/9781315271767

[88] Singh P, Patil Y, Rale V. Biosurfactant production: Emerging trends and promising strategies. Journal of Applied Microbiology. 2019;**126**(1):2-13

[89] Santos DKF, Rufino RD, Luna JM, et al. Biosurfactants: Multifunctional biomolecules of the 21st century. International Journal of Molecular Sciences. 2016;**17**:401. DOI: 10.3390/ijms17030401

[90] Natural Surfactants Market. 2018. Available from: https://www.marketsandmarkets.com/Market-Reports/natural-surfactant-market-25221394.html [Accessed: 14 August 2021]

[91] Souza EC, Vessoni-Penna TC, De Souza Oliveira RP. Biosurfactant-enhanced hydrocarbon bioremediation: An overview. International Biodeterioration and Biodegradation. 2014;**89**:88-94. DOI: 10.1016/j.ibiod.2014.01.007

[92] Geys R, Soetaert W, Van Bogaert I. Biotechnological opportunities in biosurfactant production. Current Opinion in Biotechnology. 2014;**30**:66-72

[93] Marchant R, Banat IM. Microbial biosurfactants: Challenges and opportunities for future exploitation. Trends in Biotechnology. 2012;**30**(11):558-565

[94] Hayes DG. Fatty acids-based surfactants and their uses. In: Fatty Acids. Elsevier Inc.; 2017. pp. 355-384

[95] Bhadani A, Iwabata K, Sakai K, Koura S, Sakai H, Abe M. Sustainable oleic and stearic acid based biodegradable surfactants. RSC Advances. 2017;**7**(17):10433-10442

[96] Farias CBB, Almeida FCG, Silva IA, Souza TC, Meira HM, Soares da Silva RCF, et al. Production of green

surfactants: Market prospects.
Electronic Journal of Biotechnology.
2021;**51**:28-39. DOI: 10.1016/j.
ejbt.2021.02.002

[97] Makkar RS, Cameotra SS, Banat IM.
Advances in utilization of renewable
substrates for biosurfactant production.
AMB Express. 2011;**1**(1):1-19

[98] Marion P, Bernela B, Piccirilli A,
Estrine B, Patouillard N, Guilbot J, et al.
Sustainable chemistry: How to produce
better and more from less? Green
Chemistry. 2017;**19**(21):4973-4989

[99] Perathoner S, Centi G. Science and
Technology Roadmap on Catalysis for
Europe. Brussels: ERIC aisbl; 2016

[100] Lange JP. Renewable feedstocks:
The problem of catalyst deactivation
and its mitigation. Angewandte Chemie,
International Edition.
2015;**54**(45):13187-13197

[101] Gu Y, Jérôme F. Bio-based solvents:
An emerging generation of fluids for the
design of eco-efficient processes in
catalysis and organic chemistry.
Chemical Society Reviews.
2013;**42**(24):9550-9570

[102] Jessop PG, Mercer SM,
Heldebrant DJ. CO_2-triggered switchable
solvents, surfactants, and other
materials. Energy & Environmental
Science. 2012;**5**(6):7240-7253

[103] Eckert CA, Knutson BL,
Debenedetti PG. Supercritical fluids as
solvents for chemical and materials
processing. Nature. 1996;**383**(6598):
313-318

Influence of Tween 80 Surfactant on the Binding of Roxatidine Acetate and Roxatidine Acetate–loaded Chitosan Nanoparticles to Lysozyme

Mohsen T.A. Qashqoosh, Faiza A.M. Alahdal,

Yahiya Kadaf Manea, Swaleha Zubair and Saeeda Naqvi

Abstract

The drug binding to protein is an attractive research topic. In order to assess the release of RxAc-CsNPs and their binding with lysozyme under physiological conditions, nanocomposite materials based on chitosan (Cs) and Roxatidine acetate (RxAc) in the presence Tween 80 (Tw80) surfactant were developed. The addition of Tw80 to CsNPs increased RxAc release in vitro. In this work, Stern–Volmer plot and thermodynamic results indicated that the mechanism of Lyz with RxAc and Lyz with RxAc-CsNPs was static mechanism and the main forces in both systems were hydrogen bonding and Van der Waals forces, which indicated that the binding reaction in both systems is spontaneous, exothermic and enthalpically driven. Synchronous fluorescence and CD results indicated that the RxAc and RxAc-CsNPs cause change in the secondary construction of Lyz. It was also found that the addition of Tw80 affects the binding constant of drug with protein. Finally, the molecular docking results have also been in accordance with the results of other techniques. Hence, the developed RxAc loaded Chitosan nanoparticles could be used as an effective strategy for designing and application of the antiulcer drugs. Altogether, the present study can provide an important insight for the future designing of antiulcer drugs.

Keywords: Roxatidine acetate, lysozyme, tween 80, chitosan nanoparticles, spectroscopy, molecular docking

1. Introduction

The interactions of proteins with chemicals have prompted increasing research interest in recent years. Proteins are remarkable biomolecules presenting different functions and roles. Some are specific to their biological actions whereas some are selective toward the binding site [1, 2]. Conformational changes of protein may influence its transportation, function, assembly, potential cytotoxicity, and tendency to aggregate [3, 4]. Furthermore, it has been indicated that the serum albumin conformation will be changed upon binding with ligands or molecules, and the change shows its influence on the secondary and tertiary structures of albumins and their biological function as a carrier protein [5, 6]. Some diseases (such as Alzheimer's disease, Parkinson's disease, and amyloid disease) are related to protein misfolding [6]. The thermodynamic and kinetic study of proteins plays an important role in understanding biological functions ranging from genetic information to molecular diagnostics [7, 8]. The functions and structure of a protein are strongly related to each other and due to this, protein folding/unfolding has protruded as an important property in biochemistry and biophysics [9, 10]. Therefore, the studies of such chemicals and their bindings with proteins are of fundamental and imperative importance.

The binding of nanoscale materials with proteins has become the most common with the availability of organic polymers, inorganic nanoparticles, carbon nanotubes, etc.. Recently, the nanoparticle studies have opened new avenues to study biomolecular interactions with their applications as drug delivery, biocompatibility, diagnostics, and smart materials. The binding of nanocolloidal particles to proteins has also been formed, since the study of immunoprobes in the early 1970s [11]. Moreover, various studies of peptide or protein including lysozyme binding with nanoparticles of different sizes have been conducted. In the process, the proteins, generally, suffer a significant loss in enzyme activity and a partial loss of structure [12, 13]. Thus, it should be expected that the size of the particle plays a major role in changing protein structure and function [13]. However, no systematic study has been performed to date on the effect of Roxatidine acetate–Chitosan nanoparticles on the structure and function of Lysozyme. For this reason, we embarked on a study of protein binding with Roxatidine acetate–Chitosan nanoparticles.

Lysozyme (Lyz) (**Figure 1A**) is one of the important proteins that is found in the blood and has various functions but is similar in its tendency to bind ligands/drugs. Lyz is an antibacterial and antiviral protein found in various biological tissues and fluids, such as skin, liver, lymphatic tissues, tears, saliva, and blood of human and other animals [13, 14]. Lyz is unique in its ability to hydrolyze the β-1,4 glycoside bond between N-acetylglucosamine and N-acetylmumaric acid of the gram-positive bacteria, thus protecting the body against the bacterial invasion [15]. Some of its important biological roles also include antihistaminic, anti-inflammatory, and antineoplastic activity [15–18]. Lyz consists of 129 amino acid residues and contains six tryptophan (Trp) and three tyrosine (Tyr) residues [4, 15]. Three residues of Trp

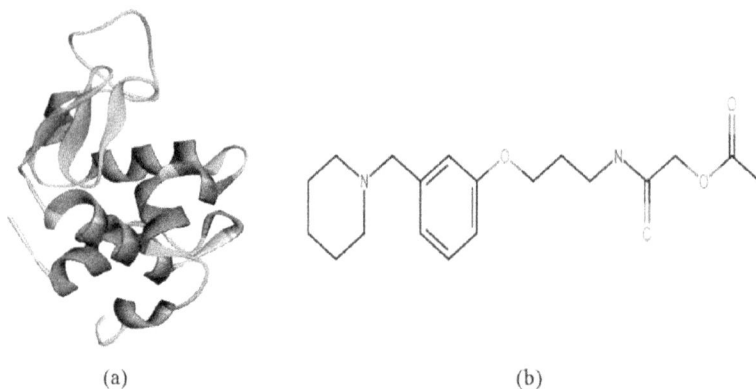

(a)   (b)

Figure 1.
(A) Three dimensional structure of Lyz, (B) chemical structure of Roxatidine acetate.

are placed at the binding sites, two sites in the hydrophobic cavity, while the last site is located independently from others [4, 15, 19–21], and the effectiveness of drugs depends on both pharmacokinetic and pharmacodynamic factors. Therefore, the studies on the interactions of drugs and Lyz are of importance in understanding the release disposition, transportation, and metabolism of drug as well as the efficacy process involving drug and Lyz. Lyz has been preferentially used as a model protein to study the protein folding/unfolding, dynamics, and ligand interaction due to its small size, abundance, high stability, and ability to bind and drug carrying capacity [22–24].

Roxatidine acetate (RxAc) (**Figure 1B**) is an antagonist of a histamine H:z-receptor, which rapidly turns into Roxatidine through esterases in the plasma, small intestine, and liver (its active metabolite) [25, 26]. Roxatidine is an active inhibitor of gastric acid secretion in humans and animals [3, 25] and does not overlap with other drugs in the hepatic metabolism and has no antiandrogenic influences as most other H:z-receptor antagonists [27]. Wide-scale experiments have illustrated that 150 mg of Roxatidine acetate per day is recommended as typical dosages of ranitidine and cimetidine in the patients for treatment of gastric ulcer or duodenal ulcer [25, 27] and that 75 mg of Roxatidine acetate as dosage in the evening is probably a standard amount for the prohibition of peptic ulcer recurrence [25, 28]. Primary studies also mention that Roxatidine acetate is perhaps useful in the treatment of stomach ulcer and reflux esophagitis and in the protection of pulmonary acid aspiration [25].

Spectroscopic techniques are mostly used to detect the accessibility of quenchers to fluorophore groups of albumin and help to understand the binding mechanism of albumin to small molecules and clarify the nature of the binding phenomenon [11, 29].

In the present study, the RxAc and RxAcNPs interactions with Lyz were methodically investigated and analyzed using diverse spectroscopic techniques to reveal the binding types and properties of RxAc and RxAcNPs with Lyz. The influences of RxAc and RxAcNPs on the conformation and microenvironment of Lyz were explored.

The aim of this study was the synthesis and characterization of Roxatidine acetate–loaded Chitosan in the presence of Tween80 (Tw80) surfactant (RxAcNPs) and to clarify the binding mechanism of RxAc and RxAcNPs with Lyz using multi-spectroscopic and molecular docking techniques and provide useful information for understanding the toxicological actions of RxAc and RxAcNPs at the molecular level.

2. Experimental

2.1 Materials

Lysozyme (from hen egg white) (Catalog number: L6876) was purchased from Sigma and was used as such. The Lysozyme solution was prepared in the 0.1 M phosphate buffer of pH = 7.40. The concentration of Lyz was determined using the extinction coefficient ϵ_{280} = 37,750 mol^{-1} L cm^{-1} [30]. Chitosan, Sodium tripolyphosphate (TPP), and Tween80 (Tw80) were also purchased from Sigma (India). NaCl (0.15 M) has been added to buffer solutions to control the ionic strength, as required. Roxatidine acetate HCl (RxAc) (≥ 98%) was purchased from Tokyo Chemical Industry Co., Ltd. (TCI), India. The stock solution of RxAc (0.3 mM) was prepared in ethanol and the final concentration of ethanol was below 2.3%, and the stock solution of RxAcNPs (0.3 mM) was also prepared in ethanol and

Components	Ratio (5:1:1)
Roxatidine acetate (mg)/100 ml	30
Chitosan (g)/100 ml	0.5
Sodium Tripolyphosphate (g)/100 ml	0.1
Tween 80 0.1 % (ml)	20
Acetic acid ml/100 ml	2

Table 1.
Formula for preparation of Roxatidine acetate loaded Tween80-Chitosan nanoparticles.

this concentration was used for all the spectroscopic measurements. RxAc and RxAcNPs were accurately weighed on Shimadzu AUY-220 microbalance of resolution 0.1 mg. All the reagents were of analytical grade. For all experiments, double-distilled water was used.

2.2 Synthesis of Roxatidine acetate nanoparticles

Roxatidine acetate drug-loaded Tween80-Chitosan nanoparticles (RxAcNPs) were prepared through the ionotropic gelation technique. The principle of this method is interaction of the positive charge of Chitosan amino groups with the negative charge of TPP groups [31, 32]. As listed in **Table 1**, the solution of Chitosan was prepared by dissolving 0.5 g Chitosan (0.5%) in 100 ml of acetic acid (1% v/v). TPP solution (0.1%) was prepared through dissolving 100 mg of TPP in 100 ml of deionized water. 30 mg of Roxatidine acetate was added to the solution of TPP. The solution was stirred at 1500 rpm for 30 min using an ultrasonicator (vibronics), and the solution of TPP was added gradually with continuous stirring for 3 hours on a homogenizer. The mixture of Roxatidine acetate, 0.1% TPP, and 0.1% Tween80 was added gradually to Chitosan solution. Tween80 was added to make the prepared solution stabilized and to limit the nanoparticle growth and thus to obtain particles of reduced mean sizes [31–33]. The precipitate was stirred at 9000 rpm for 3 hours using an ultrasonicator (vibronics). After the addition of a drug–Tween80–TPP solution to the solution of Chitosan, the suspended solution of nanoparticles was centrifuged for 15 min at 10,000 rpm, and the Roxatidine acetate nanoparticles were obtained.

2.3 Drug content and release profile of Roxatidine acetate nanoparticles

To confirm the drug content, encapsulation efficiency, and release of RxAc, the conventional method and dialysis method were used for testing RxAc-loaded Tween80-CsNPs. The encapsulation efficiency and drug content were estimated according to the procedure reported by Cevher et al. [34]. After drug loading, the RxAcNPs were isolated from the suspension using centrifugation at 10000 rpm for 15 min. The quantity of free Roxatidine acetate in the supernatant was measured using the UV–Vis spectrophotometer (double beam Perkin Elmer λ-45) at 275 nm. RxAc release was investigated in vitro by dialysis using phosphate buffer (PBS) at different pH (3.5, 6.6, 7.4, and 8.4) and 298 K. 25 mg of Roxatidine acetate–loaded Tween80–Chitosan nanoparticles were added to 50 ml of each PBS buffer in different flasks and were shaken using a magnetic stirrer at 298 K. At various time intervals, 2 ml from the suspended solution of nanoparticles was taken and centrifuged at 10000 rpm for 15 min and the standard curve for RxAc was acquired by UV spectrophotometry. At 275 nm, the RxAcNPs entrapment efficient (EE), drug content, and accumulated release percentage (%) at different pH were determined spectrophotometrically and were calculated using the following equations, as described previously [31–35]:

$$drug \ content \ (\%) = \frac{weight \ of \ drug \ in \ nanoparticles}{weight \ of \ nanoparticles} \times 100 \qquad (1)$$

$$EE \ (\%) = \frac{(Amount \ of \ drug \ taken \ for \ formulation) - (amount \ of \ unentrapted \ drug)}{Total \ amount \ of \ drug \ in \ formulation}$$

$$\times \ 100$$

$$(2)$$

2.4 Characterization of Roxatidine acetate–loaded Tween80–Chitosan nanoparticles

Characterization of RxAc-loaded Tween80–Chitosan nanoparticles comprised Fourier transform infrared spectroscopy using PerkinElmer Frontier equipment with a resolution of 4 cm^{-1} and a wavenumber range of 400–4000 cm^{-1}. The particle size was measured by Zetasizer Nano ZS (Malvern, UK), and scanning electron microscope (SEM) in a JEOL JSM-6510 with an accelerating voltage of 15 kV was used to visualize the shape of RxAcNPs. Powder XRD scanning (Lab-X, Shimadzu-XRD 6100 instrument, Japan) was performed to analyze the crystalline nature of RxAcNPs within the range of diffraction angle 2θ from 5° to 60.

2.5 Analysis of RxAc and RxAcNPs with Lyz

2.5.1 Fluorescence spectra study

The spectra of fluorescence emission were collected on Hitachi F-2700 Spectrofluorimeter with a Xenon lamp, and the quartz cuvette of 1 cm path length was used. The slit widths of excitation and emission were set at 5 nm. The rate of scanning was set to 300 nm/min. The wavelength of excitation was set at 280 nm and emission wavelength at 290–500 nm. The synchronous fluorescence spectra were scanned from 260 to 330 nm (Δλ = 15 nm) and from 220 to 330 nm (Δλ = 60 nm). A buffer blank spectrum was subtracted from the measured spectra for fluorescence background correction. The concentration of Lyz was kept constant at 10 μM, while the concentrations of RxAc and RxAcNPs were varied. All the measurements were performed at pH 7.4.

2.5.2 The influence of Tween80 (Tw80) inclusion on the interaction of the Lyz–RxAc system

Tween80 was utilized to improve the stability of the therapeutic molecule and its safety at its target site. The influence of Tw80 inclusion on the interaction of the Lyz– RxAc system was studied by keeping the concentration of Lyz at 10 μM and changing the concentration of RxAc (2–16 μM), while the concentrations of Tw80 were maintained at 2 and 4 μM.

2.5.3 UV–Vis spectroscopic measurements

The UV–Vis absorption spectra were recorded using a double-beam PerkinElmer λ-45 spectrophotometer. For the whole experiment, the quartz cuvette of 1 cm path length was used. The concentration of Lyz was kept at 10 μM while the RxAc and RxAcNPs concentrations were varied. All the readings were recorded at room temperature.

2.5.4 Circular dichroism (CD) measurements

Circular Dichroism (CD) spectra were carried out using a Jasco J-815 spectropolarimeter and using a quartz cell of 0.1 cm path length. Response time and data pitch were fixed at 1 s and 1 nm, respectively. CD spectra were measured in the far-UV region (200–250 nm) with a scan speed of 100 nm/min and two scans for each spectrum under constant nitrogen flow. For all the measured spectra, Phosphate buffer baseline subtraction (pH 7.4) was used. Concentration of Lyz for all runs was fixed at 10 μM, while the RxAc and RxAcNPs concentrations were 0, 40, and 80 μM. All the measurements were carried out at room temperature.

2.5.5 Molecular docking of the Lyz–RxAc system

Molecular docking study used software Autodock 4.2 and Autodock tools (ADT) using the Lamarckian genetic algorithm [29]. The crystal structure of Lyz (PDB ID: 2LYZ) was obtained from Brookhaven Protein Data Bank and three-dimensional structure of Roxatidine acetate (CID = 5105) was obtained from PubChem. All the ions and water molecules were removed, hydrogen atoms were added, and partial Kollman charges were assigned. The Autodock run was carried out through the following parameters: GA population size, 150; maximum number of energy evolutions, 2.5×106, and Grid box size 86 Å × 80 Å × 96 Å along x-, y-, and z axes covering the whole protein with a grid-point spacing of 0.375 Å. Discovery Studio 3.5 was utilized for identification and visualization of the residues involved.

3. Results and discussion

3.1 Method development and release profile

In order to obtain insight into the binding interaction of RxAcNPs with Lyz, the drug content, encapsulation efficiency (EE), and releasing percentage of RxAc were determined utilizing spectrophotometric techniques, with the mole ratio 5:1:1 of Cs, TPP, and Tween80, respectively. The maximum absorption wavelength was found to be 275 nm for RxAc. The drug content and encapsulation efficiency of RxAc in CsNPs based on the preparation of formulation are represented in **Table 2**. As listed in **Table 2**, the results displayed high encapsulation efficiency, which was 88.25 ± 0.26%, and the total content of drug in the nanoform was 26.48 ± 0.17 mg, which was nearly the total content of drug used in the preparation of the formulation matrix. These results revealed that the developed method is reliable and accurate to estimate the content of drug without interference of the formulation matrix or excipients. Additionally, it has the possibility to estimate the content of drug in the complex nanocarriers-based formulation.

The RxAc release profile from RxAc-loaded Tween80–Chitosan nanoparticles at different values of pH is illustrated in **Figure 2**. The drug released from the nanoparticles was little during the initial 2 hours (less than 25%). After 2 hours, the quantity of the released drug increased with time. The RxAc percentages released at the end of 24 hours. were 90.21 ± 0.73, 85.83 ± 0.54, 82.79 ± 0.34, and 75.01 ± 0.57% for pH 3.5, 6.6, 7.4, and 8.4, respectively (**Table 3** and **Figure 2**). Furthermore, as the pH decreased, the amount of released drug increased, showing

The total quantity of Roxatidine acetate used in formulation (mg)	30.00
The cumulative quantity of RxAc (mg $\pm$ SD*)	26.48 $\pm$ 0.17
Encapsulation efficiency (% $\pm$ SD*)	88.25 $\pm$ 0.26
Particle Size (nm $\pm$ SD*)	220 $\pm$ 5
*Standard deviation (N = 3)	

Table 2.
Nanoparticle sizes and mass balance of the Roxatidine acetate used in nanoparticles formulation.

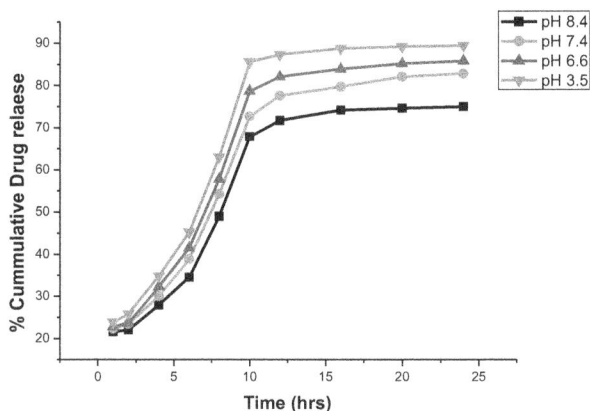

Figure 2.
In vitro of RxAcNPs release profile at different pH values.

that the drug release depends upon the pH of the media, as well as the nature of the polymer matrix [33, 34], which means that the developed method is suitable and effective for preparing the antiulcer drugs in nanoform.

The nanoparticles resulting from this developed method were used to investigate the applicability in simulation of studies of drug nanoparticles–protein interaction. The known concentrations (0–16 μM) of the RxAcNPs solution were added to the fixed concentration of Lyz (10 μM) to examine the binding interaction under physiological conditions.

3.2 Characterization of Roxatidine acetate–loaded Tween80–chitosan nanoparticles

Fourier transform infrared (FTIR) spectra of the CsNPs and RxAcNPs are shown in **Figure 3A**. Chitosan is known to possess amine groups on the glucosamine moiety whereas Roxatidine acetate is an amphoteric drug having hydrophobic and hydrophilic moieties (—OH, CO, and —NH groups). The characteristic absorption bands for Chitosan were observed at 1650, 1545 and 1420 cm^{-1} were corresponding to amide I, amide II and amide III, respectively. 1095 cm^{-1} was corresponding to C—N stretching, and 2936 cm^{-1} was corresponding to the asymmetric stretching vibration of methylene and 3350 cm^{-1} was due to the stretching vibration of N—H. The FTIR spectra of RxAcNPs were compared with the FTIR spectra of CsNPs. The spectra did not show any new band for characteristic peaks of RxAc in RxAcNPs spectra and the existing shift of bands indicating entrapment of RxAc within the chitosan matrix, suggesting no new chemical bond formation between RxAc and CsNPs. Consequently, this observation excluded the possibility of an interaction between the polymer and drug indicating

pH Amount (mg) ± SD*	3.5	6.6	7.4	8.4
The total amount of Roxatidine acetate used for release study	25	25	25	25
The cumulative amount of drug released	20.32 ± 0.11	18.49 ± 0.39	16.82 ± 0.29	14.49 ± 0.46
Amount of drug unreleased	2.23 ± 0.24	2.97 ± 0.41	3.88 ± 0.18	4.26 ± 0.27
The total amount of drug recovered	22.55 ± 0.49	21.46 ± 0.05	20.70 ± 0.16	18.75 ± 0.21
Percentage amount of drug recovered (%)	90.21 ± 0.73	85.83 ± 0.54	82.79 ± 0.34	75.01 ± 0.57

*Standard deviation (N = 3).

Table 3.
Mass balance of Roxatidine acetate used in vitro release study at deferent pH.

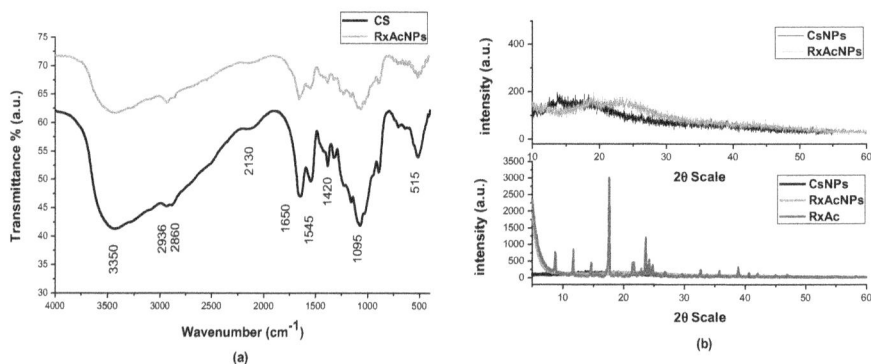

Figure 3.
(a) FTIR spectra for Roxatidine acetate (RxAc), Chitosan nanoparticles (CsNPs), Roxatidine acetate loaded Chitosan nanoparticles (RxAcNPs), (b) PXRD spectra for Roxatidine acetate (RxAc), Chitosan nanoparticles (CsNPs), Roxatidine acetate loaded Chitosan nanoparticles (RxAcNPs).

that RxAc was physically dispersed in the polymer [33–38]. As shown in **Figure 3B**, RxAc-loaded Tween80–Chitosan nanoparticles were examined using the PXRD technique. The peak at 11.72° represents the presence of Cs and at 17.65°, the presence of TPP is indicated. A synthesized nanoform was specified, illustrating the semicrystalline nature of RxAcNPs after the available analysis, which depends on the little sharp pattern of XRD; hence, the drug was just encapsulated in Tween80–Cs nanoparticles without any interaction. The peaks at 24.18° and 27.21° indicated the presence of RxAc [39–45]. As shown in **Figure 4A** and **Table 2**, the data of DLS showed that the particle size of RxAcNPs was 220 ± 5 nm, which was almost in conformity with the data of SEM as shown in **Figure 4B**. The SEM micrograph of RxAcNPs clearly illustrates the presence of RxAc on the Chitosan surface, which clarifies the drug encapsulation in the nanoparticle surface. The data of SEM of Tw80–CsNPs before and after loading RxAc illustrated that the spherical shape of the nanoparticles of Tw80–CsNPs is slightly deformed because of the loading of RxAc, as shown in **Figure 4B**.

3.3 Analysis of RxAc and RxAcNPs with Lyz

3.3.1 Fluorescence spectroscopy

Fluorescence quenching of lysozyme is broadly used in measuring the binding affinity of protein and drug. Lyz has three main Trp residues located at its active

(a)

(b)

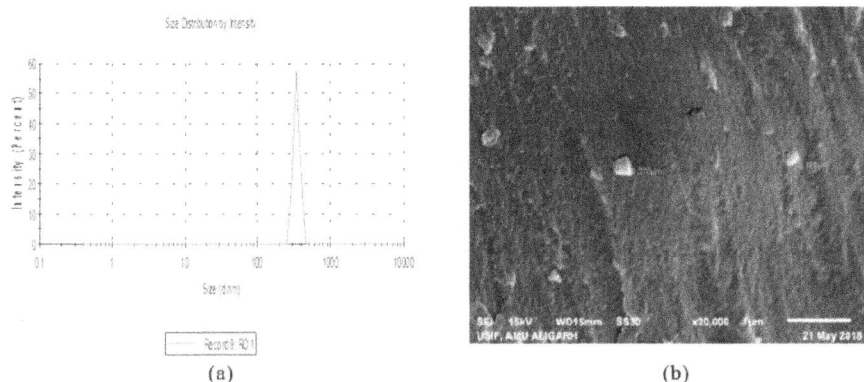

Figure 4.
(a) Particle size distribution of RxAcNPs, (b) SEM image of RxAc-loaded Chitosan nanoparticles (RxAcNPs).

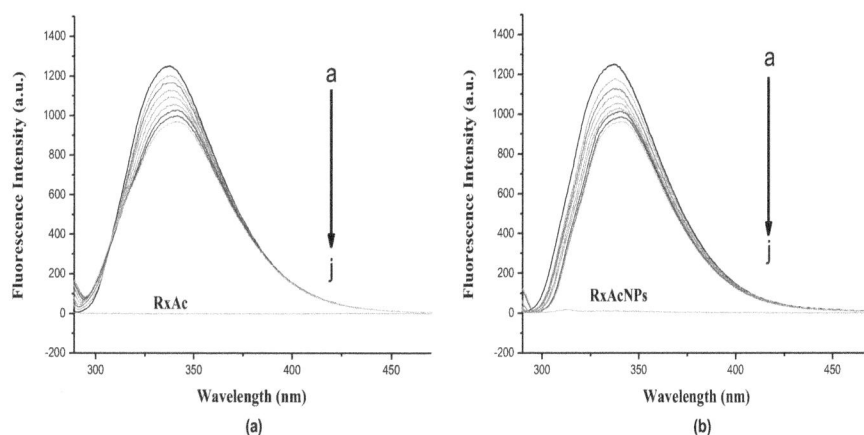

(a)

(b)

Figure 5.
Fluorescence emission spectra of Lyz in the presence of (a) RxAc and (b) RxAcNPs at 298 K. C_{Lyz} : 10μM (a), $C_{RxAc\ or\ RxAcNPs}$ (b-i): 2, 4, 6, 8, 10, 12, 14 and 16 μM; native RxAc or RxAcNPs (j): 2 μM.

binding site, i.e., Trp-62, Trp-63, and Trp-108. The intrinsic fluorescence of Lyz comes from tryptophan residues (Trp-62, Trp-63, and Trp-108) and to study the conformational changes of Lyz in the binding process of Lyz with drugs used to be a fluorescent probe [4, 46]. The effects of RxAc and RxAcNPs on Lyz fluorescence intensity are shown in **Figure 5A** and **B**, respectively. After being excited with a wavelength of 280 nm, Lyz has a fluorescence emission with a peak at 337 nm; the fluorescent intensity of Lyz decreased regularly with increasing concentrations of RxAc and RxAcNPs. Interestingly, a red shift of about 6 nm and 4 nm in the λmax were observed in Lyz-RxAc and Lyz-RxAcNPs systems, respectively. Moreover, 78% of the fluorescence emission was quenched by RxAc in case of the Lyz–RxAc system whereas 77% quenching of the emission was observed in case of the Lyz–RxAcNPs system, which sketches a picture as to how the quencher RxAc and RxAcNPs ingress the fluorophore and bring about the quenching. Further, it suggests a change in the surrounding environment of the fluorophores due to interaction with RxAc and RxAcNPs and that the binding regions of RxAc and RxAcNPs are in the vicinity of Trp residues. Considering the above observations, it could be adjudged that RxAc and RxAcNPs bind to Lyz and quench its intrinsic fluorescence. The red shift in the λmax in Lyz–RxAc and Lyz–RxAcNPs systems indicated an increase in polarity and a decrease in hydrophobicity [47–49].

As we know well, the phenomena of fluorescence quenching are brought about by various intermolecular episodes, namely excited-state reactions, ground-state complex formation, energy-transfer molecular rearrangements, and collisional quenching [50–52]. There are two types of quenching that are Static quenching and dynamic quenching. In static quenching, a nonfluorescent fluorophore-quencher complex is formed, whereas in dynamic quenching, collision between the quencher and fluorophore during the lifetime of the excited state is established. The two types of quenching can be distinguished from each other by taking viscosity and temperature-dependent measurements [53]. In the present systems, the fluorescence-quenching mechanism has been studied using the well-known Stern–Volmer (S–V) Equation [48, 53, 54]:

$$\frac{F_0}{F} = 1 + K_{sv}[Q] = 1 + K_q \tau_0 [Q] \tag{3}$$

where F_0 and F are the protein fluorescence intensities in the absence and in the presence of the drug molecule (quencher), respectively, Ksv is the constant of Stern–Volmer quenching, [Q] is the concentration of the quencher, K_q is the quenching rate constant of the biomolecule, and τ_0 is the biomolecule average lifetime in absence of the quencher. A single type of quenching mechanism, either static or dynamic quenching mechanism, is included, when the plot of F_0/F vs. [Q] is linear, whereas deviation from linearity suggests the presence of both quenching mechanisms. The value of Ksv is estimated from the plot of F_0/F vs. [Q]. Considering the well-known connection between the quenching constant and the K_q quenching rate constant, and taking into account the fluorescence lifetime of the biopolymer as 10^{-8} s, the K_q values can be calculated [9, 19, 55]:

$$K_q = \frac{K_{sv}}{\tau_0} \tag{4}$$

Figure 6A and **B** show the plots of F_0/F for Lyz versus [Q] of RxAc and RxAcNPs at 298, 304, and 310 K and pH 7.4, where [Q] ranges from 2 to 16 μM of RxAc and RxAcNPs. Plots in **Figure 6A** and **B** show that the results of Lyz–RxAc and Lyz–RxAcNPs systems agree very well with the Stern–Volmer equation, which indicates that a single type of quenching mechanism is involved, either static or dynamic [56–59]. The results listed in **Table 4** showed that K_{SV} and K_q values of Lyz–RxAc and Lyz–RxAcNPs decreased upon increasing temperature and that the quenching of both systems follows the static quenching mechanism [53, 60]. The maximum scatter collision quenching constant (K_q) with the biopolymer is 2×10^{10} L mol^{-1} s^{-1}. The values of K_q of the protein quenching initiated by RxAc and RxAcNPs are greater than the constant of maximum scatter collision quenching, thus indicating that quenching is initiated from the formation of complex and not the dynamic collision [53].

3.3.2 Binding interaction analysis

The constant of binding (K_a) and the number of binding sites (n) of the inter-action between RxAc/RxAcNPs and Lyz could be investigated from the logarithmic form of the Stern–Volmer equation: [48, 52, 53]

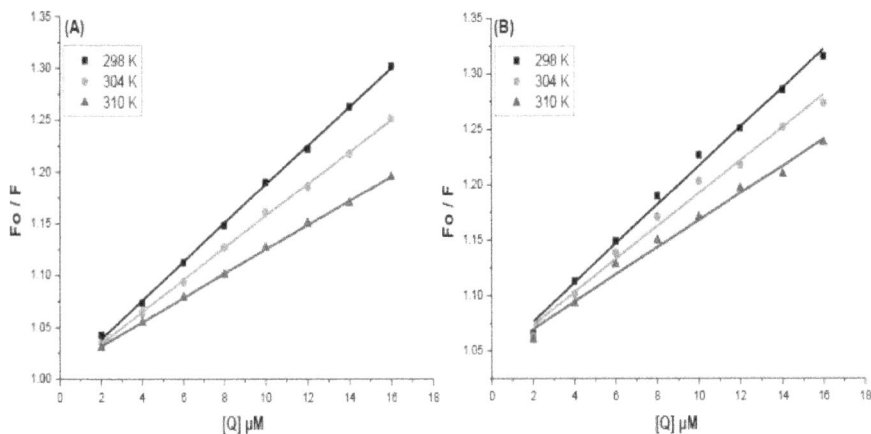

Figure 6.
Stern-Volmer plots for quenching of Lyz fluorescence by (A) RxAc (B) RxAcNPs at different temperatures.

System	pH	T(K)	$K_{SV} \times 10^4$ (L mol^{-1})	SD*	R*	$K_q \times 10^{12}$ (L mol^{-1} s^{-1})
Lyz-RxAc	7.4	298	1.86	0.04	0.999	1.86
		304	1.54	0.12	0.999	1.54
		310	1.17	0.28	0.999	1.17
Lyz-RxAcNPs	7.4	298	1.76	0.17	0.996	1.76
		304	1.49	0.13	0.994	1.49
		310	1.22	0.07	0.994	1.22
Lyz-RxAc-Tw80 (2μM)	7.4	298	1.47	0.06	0.999	1.46
Lyz-RxAc-Tw80 (4μM)	7.4	298	1.33	0.08	0.998	1.33

S.D* is standard deviation (N = 3)
R** is the correlation coefficient of K_{SV}

Table 4.
Quenching parameters of Lyz-RxAc and Lyz-RxAcNPs systems at different temperatures.

$$\log \frac{F_0 - F}{F} = \log K_a + n \, \log [Q] \qquad (5)$$

From the plot of Log[$(F_0 - F)/F$] vs. log [Q], the binding constant (K_a) and the number of binding sites (n) could be obtained, where the intercept yields the value of the binding constant (K_a) and the slope gives the number of binding sites (n) (listed in **Table 5**). The values of K_a were 10^4 L mol^{-1} for Lyz–RxAc (**Figure 7A**) indicating a high affinity of the Lyz molecule for RxAc besides binding number up to 0.96; however, the binding affinity of Lyz for RxAcNPs (**Figure 7B**) was found lower, ranging up to the order of 10^3 L mol^{-1} and binding number up to 0.91. All these results lead to the conclusion that binding is stronger between Lyz and RxAc than that between Lyz and RxAcNPs, which will definitely affect its free concentration and its bound concentration in the blood plasma [61, 62].

The drug bioavailabilities could be estimated from the binding affinity values. The nanoform of drug (RxAcNPs) has shown less binding affinity to Lyz, which

System	pH	Temp. (K)	K_a (L mol^{-1})	R*	n	ΔG (kJ mol^{-1})	ΔH (kJ mol^{-1})	ΔS (J mol^{-1} K^{-1})
Lyz-RxAc	7.4	298	1.14×10^4	0.998	0.96	−22.25	−71.80	−166.28
		304	0.84×10^4	0.999	0.95	−21.24		
		310	0.37×10^4	0.999	0.89	−20.25		
Lyz-RxAcNPs	7.4	298	1.24×10^3	0.999	0.91	−17.73	−86.61	−231.13
		304	0.69×10^3	0.999	0.88	−16.34		
		310	0.32×10^3	0.999	0.84	−14.96		
Lyz-RxAc-Tw80(2 µM)	7.4	298	2.33×10^3	0.999	0.90	-	-	-
Lyz-RxAc-Tw80(4 µM)	7.4	298	1.07×10^3	0.997	0.85	-	-	-

*R is the correlation coefficient of K_a

Table 5.
Binding constant, number of binding sites and Thermodynamic parameters of Lyz-RxAc and Lyz-RxAcNPs systems at different temperatures.

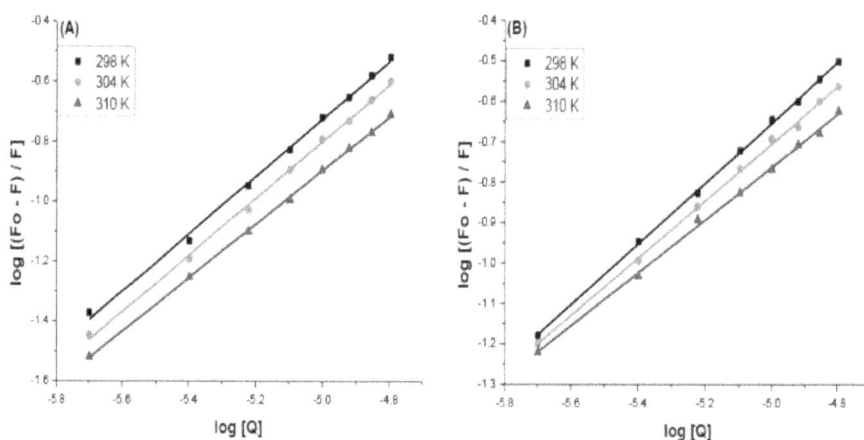

Figure 7.
Plots of log [(Fo-F)/F] versus log [Q] for (A) Lyz-RxAc and (B) Lyz RxAcNPs systems at different temperatures.

indicates that the distribution and absorption of drug nanoparticles to various tissues will be higher, as the stability of the Lyz–RxAcNPs complex is lower compared to Lyz–RxAc complex [63, 64].

3.3.3 The influence of Tween80 (Tw80) inclusion on the interaction of the Lyz–RxAc system

The influence of Tween80 inclusion onto the interaction of Lyz–RxAc systems was studied by introducing Tween80 to the Lyz–RxAc system at room temperature (**Figure 8A** and **B**). The results of the Stern–Volmer constant (K_{SV}) and binding constant (K_a) are shown in **Figures 9A,B** and **10A,B** and listed in **Tables 4** and **5**. We observed that the results of K_{SV} and K_a in the presence of Tw80 were smaller than in its absence. These results indicated that the Tw80 helps to release RxAc from the Chitosan nanoparticles, due to a fraction of RxAc binding to it by weak

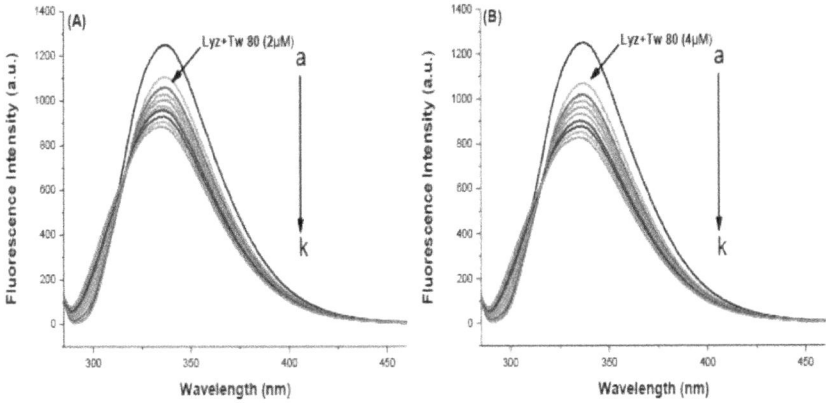

Figure 8.
A,B: Fluorescence emission spectra of Lyz in the presence of RxAc and Tw80 at 298 K. CLyz: 10 μM (a), CTw80: 2 and 4 μM (b), CRxAc (c–k): 2, 4, 6, 8, 10, 12, 14 and 16 μM.

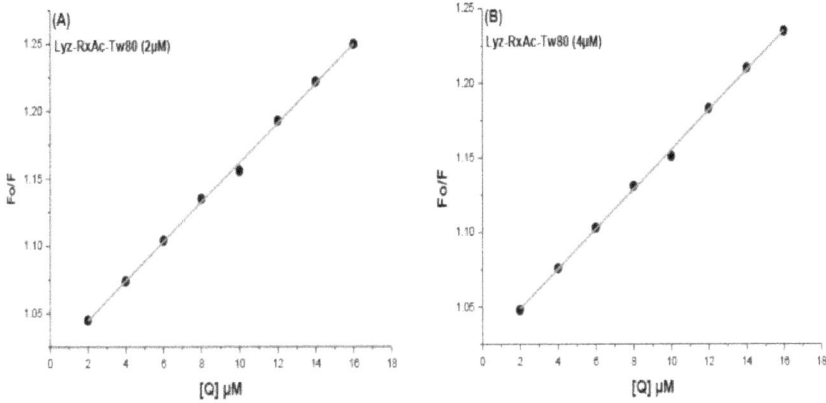

Figure 9.
A,B: Stern-Volmer plots for quenching of Lyz fluorescence by RxAc in the presence of Tw80 (2 and 4 μM) at 298 K.

bonds; hence, Tw80 helps to release more drug to the tissues as compared to the drug released from plasma [65]. Furthermore, Tw80 encloses the RxAc molecule and obstructs it from colliding directly with the amino acid residues found in the binding sites of Lyz [66].

3.3.4 The force acting between Lyz and RxAc/RxAcNPs

The driving force of binding could be assessed from the thermodynamic law summarized by Ross and Subramanian. The stability of the protein–drug complex and the binding of drug onto protein are influenced by various types of noncovalent forces such as hydrophobic interactions, hydrogen binding, Van der Waals, and electrostatic forces. To get the thermodynamic parameters, the Van't Hoff equation has been used:

$$ln \, K_a = \frac{-\Delta H^0}{RT} + \frac{\Delta S^0}{R} \tag{6}$$

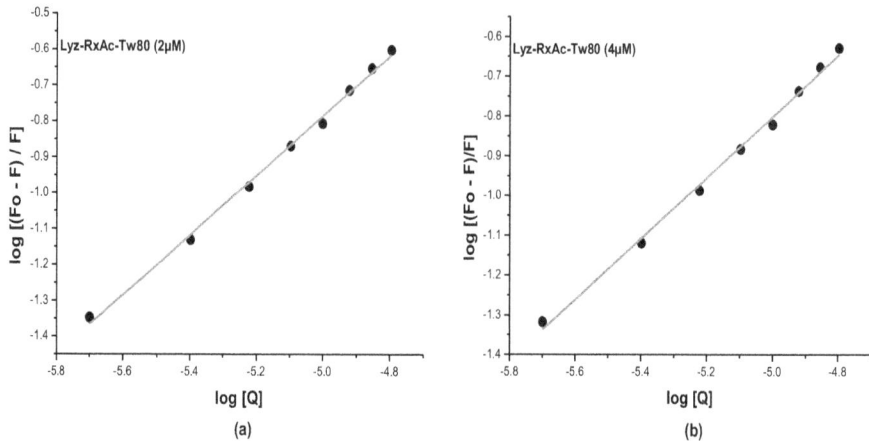

Figure 10.
A,B: Plots of log [(Fo-F)/F] versus log[Q] for Lyz-RxAc-Tw80 systems.

$$\Delta G^0 = -RTlnK_a = \Delta H^0 - T\Delta S^0 \qquad (7)$$

where K_a is the constant of binding at the corresponding temperature T, T is the absolute temperature, and R is the universal gas constant. The plot of lnK_a versus $1/T$ allows the estimation of the enthalpy change (ΔH) and the entropy change (ΔS) [9, 67–69]. The enthalpy change (ΔH) and the entropy change (ΔS) can be obtained from the slope and the intercept of the Van't Hoff plots, respectively. From the thermodynamic viewpoint, Ross and Subramanian recommended that $\Delta H < 0$ and $\Delta S < 0$ suggest the van der Waals force and hydrogen bond formation, $\Delta H > 0$ and $\Delta S > 0$ show a hydrophobic interaction, and $\Delta H < 0$ and $\Delta S > 0$ propose electrostatic forces of interaction [53, 62–64].

As shown in **Figure 11A** and **B**, there is a good linear relationship between lnK_a and $1/T$, suggesting that ΔH is constant in the current temperature range. From **Table 5**, it could be seen that $\Delta H = -71.80$ kJ mol^{-1} and $\Delta S = -166.28$ J mol^{-1} K^{-1} for the Lyz–RxAc system and $\Delta H = -86.61$ kJ mol^{-1} K^{-1} and $\Delta S = -231.13$ J mol^{-1} K^{-1} for the Lyz–RxAcNPs system. The negative values of ΔH and ΔS for interaction of Lyz with RxAc and Lyz with RxAcNPs indicate that hydrogen bonds and Van der Waals forces play a major role in the interaction of Lyz–RxAc and Lyz–RxAcNPs systems, and the binding reaction is exothermic and enthalpically driven. The negative values of ΔG for both systems at different temperatures (298, 304, and 310 K) mean that the binding processes are spontaneous in both systems.

3.3.5 Fluorescence resonance energy transfer (FRET)

Fluorescence resonance energy transfer is a nondestructive spectroscopic method and an investigatory tool that can monitor the proximity and relative angular orientation to study energy transfer from donor to acceptor. A transfer of energy could be carried out through direct electrodynamic interaction between the primarily excited molecule and its neighbors [70, 71]. The fluorophores of donor and acceptor can be entirely nonattached or attached to the same macromolecule [72]. In the present case, Lyz is the donor and RxAc and RxAcNPs are the acceptors. According to this theory, the efficiency (E) of energy transfer from Lyz to RxAc or RxAcNPs and the distance (r) of binding between Lyz and RxAc or RxAcNPs could be calculated by Eq. (8) [70, 73]:

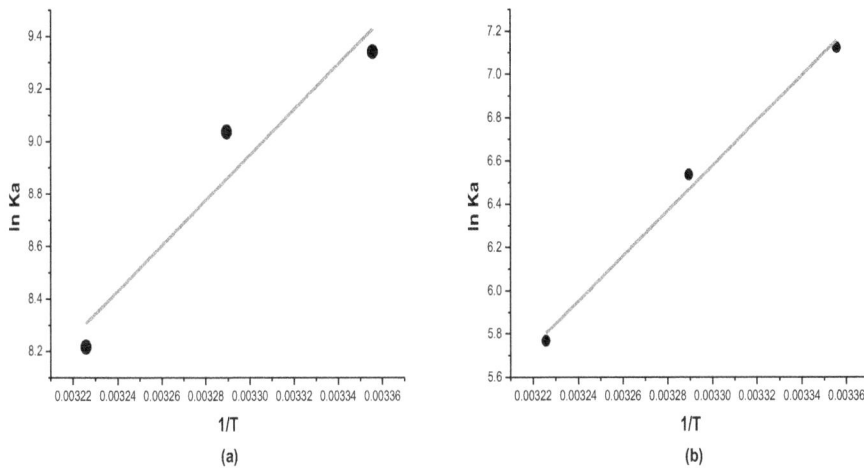

Figure 11.
Vant-Huff Plot for (A) Lyz-RxAc and (B) Lyz-RxAcNPs systems at different temperatures.

$$E = \frac{R_0^6}{R_0^6 + r^6} = 1 - \frac{F}{F_0} \tag{8}$$

where E could be determined experimentally from the donor emission in the absence (F_0) and presence of the acceptor (F), normalized to the same donor concentration, r is the actual distance between the donor (Lyz) and the acceptor (RxAc/RxAcNPs), R_0 is the critical distance when the efficiency of transfer is 50%, which depends on the quantum yield of the donor, the extinction coefficient of the acceptor, the overlap of donor emission and acceptor absorption spectra, and the mutual orientation of the chromophores. R_0 can be defined by Eq. (9) [70, 74]:

$$R_0^6 = 8.8 \times 10^{-25} K^2 N^{-4} \Phi J \tag{9}$$

where K^2 is the spatial factor of orientation related to the geometry of the donor and acceptor of dipoles, N is the refractive index of the medium, Φ is the fluorescence quantum yield of the donor, and J is the effect of spectral overlap between the donor emission spectrum and the acceptor absorption spectrum, which could be calculated by Eq. (10)

$$J = \frac{\sum F(\lambda) \, \varepsilon(\lambda) \lambda^4 \Delta \lambda}{F(\lambda) \wedge \lambda} \tag{10}$$

where $F(\lambda)$ is the donor fluorescence intensity at wavelength λ and $\varepsilon(\lambda)$ is the molar absorption coefficient of the acceptor at wavelength λ. The efficiency of transfer (E) could be obtained using Eq.8, where F and F0 are the relative fluorescence intensities in the presence and absence of acceptor [56]. For Lyz, $K^2 = 2/3$, N = 1.36, and Φ = 0.15 [62, 75].

The overlap of the absorption spectrum of RxAc and RxAcNPs with the fluorescence emission spectrum of Lyz are shown in **Figure 12A and B**, in the wavelength range of 280–310 nm and 280–308 nm, respectively. The fluorescence emission from both systems at an excitation wavelength of 280 nm is mainly from the Lyz molecule as both RxAc and RxAcNPs are nonfluorescent at the excitation wavelength. However, at this excitation wavelength, RxAc and RxAcNPs do show weak

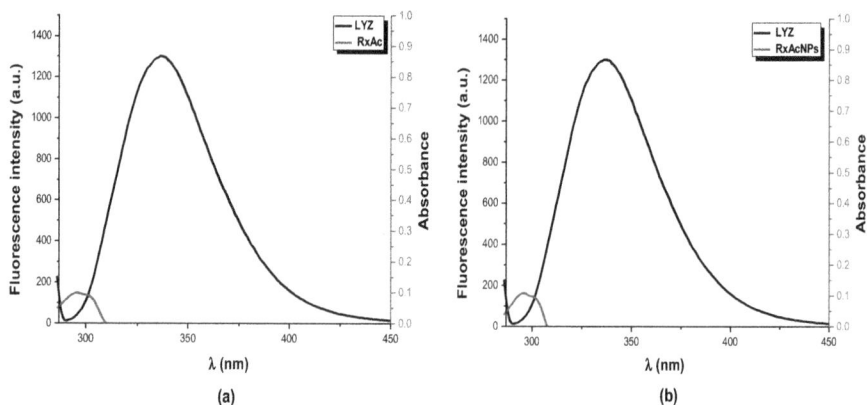

Figure 12.
Spectral overlap between fluorescence emission spectrum of Lyz and absorption spectrum of (A) RxAc and (B) RxAcNPs when the molar ratio of Lyz and RxAc or RxAcNPs is 1:1. [Lyz]: 10 μM, [RxAc or RxAcNPs]: 2 μM at 298 K.

System	R_0 (nm)	r (nm)	E	J (cm^3 L mol^{-1})
Lyz-RxAc	3.40	4.66	0.23	6.67×10^{-14}
Lyz-RxAcNPs	3.44	4.53	0.24	7.18×10^{-14}

Table 6.
Energy transfer parameters for Lyz-RxAc and Lyz-RxAcNPs interactions at 298 K.

absorption, which suggests the probability of energy transfer from Lyz to RxAc/RxAcNPs. Using Eqs. (8)–(10), the parameters related to energy transfer from Lys to RxAc or RxAcNPs are calculated and are presented in **Table 6**. The values of R_0, r, J, and E were found to be 3.40 nm and 4.66 nm, 6.67×10^{-14} cm^3 L mol^{-1} and 0.23 for Lyz–RxAc, whereas the corresponding values were 3.44 nm, 4.53 nm, 7.18×10^{-14} cm^3 L mol^{-1} and 0.24 for Lyz–RxAcNPs, respectively. The obtained result indicates that RxAc and RxAcNPs are strong quenchers and these may situate in the close proximity of Lyz. The values of binding distance (r) between the donor and acceptor for all the systems are in the range of 2–7 nm, denoting that the energy transfer is possible between Lys and RxAc or RxAcNPs. The values of R_0 and r are also in the academic range, which proves that nonradiative energy transfer occurs between Lyz and RxAc/RxAcNPs. Furthermore, the results also suggest that static quenching is responsible for the quenching of fluorescence emission as the binding involved energy transfer from Lyz to RxAc/RxAcNPs [48, 59, 76, 77].

3.3.6 Conformational changes of lysozyme

Synchronous fluorescence spectroscopy is a kind of important method and a proficient technique, which is utilized to evaluate the conformational changes and provides the information regarding the molecular environment in the vicinage of the chromophore molecule [78, 79]. Because of its sensitivity, spectral simplification, spectral bandwidth reduction, and shunning of different perturbing effects, it can be used as an ideal and a very useful method to study the microenvironment of Trp residues by measuring the possible shift in wavelength emission maximum (λem) [48, 80]. The polarity changes around the chromophore molecule, i.e., the Lyz conformation, may be due to the shift in the position of emission maximum.

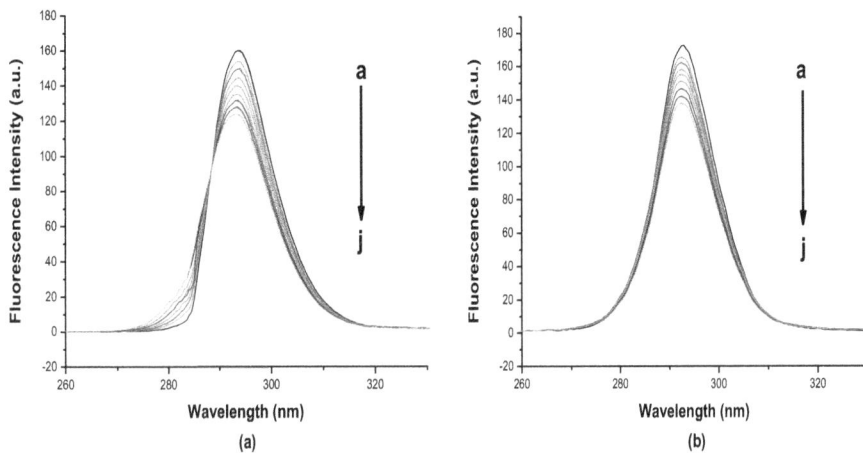

Figure 13.
*Synchronous fluorescence spectrum of (A) Lyz-RxAc and (B) Lyz-RxAcNPs systems at 298 K : (Δλ = 15 nm),
$C_{(Lyz)}$ = 10 μM; $C_{(RxAc \ or \ RxAcNPs)}$ (b-j): 2, 4, 6, 8, 10, 12, 14 and 16 μM.*

As is well-known, the spectra of synchronous fluorescence show Trp residues of Lyz
at the wavelength interval (Δλ) of 60 nm, while at the wavelength interval (Δλ) of
15 nm, the spectra of synchronous fluorescence show Tyr residues of Lyz [81].

At Δλ = 15 nm, in the Lyz–RxAc and Lyz–RxAcNPs systems in the investigated
concentration range, the maximum emission wavelength keeps its position without
any shift (**Figure 13A** and **B**), which indicates that there is no change in the
microenvironment of the Tyrosine residues in both systems, whereas over the
investigated concentration range at Δλ = 60 nm, it can be seen that the maximum
emission wavelength moderately shifts from 279 to 274 nm in the Lyz–RxAc system
and from 279 to 275 nm in the Lyz–RxAcNPs system toward blue wavelengths. On
looking through the synchronous spectra for the Lyz–RxAc and Lyz–RxAcNPs
systems (**Figure 14A** and **B**), the shift effect shows that the conformation of Lyz
has changed. The blue-shift effect indicates that the microenvironment around the
Tryptophan residues is disturbed and shows a decrease in the polarity and an
increase in the hydrophobicity around Tryptophan residues.

3.3.7 UV–vis absorbance spectroscopy

UV–Vis spectroscopy is a simple technique and an effective method that can
help to know the structural changes in the system and to explore the formation of
complex and the change in hydrophobicity [82]
In the present study, we have observed the change in the UV absorption spectra
of Lyz–RxAc and Lyz–RxAcNPs systems (**Figure 15A** and **B**), which indicated that
the interaction between Lyz and RxAc/RxAcNP molecules may lead to change in
the conformation of Lyz. It was evident that the UV absorption intensity of Lyz
increased regularly with the variation of RxAc and RxAcNP concentrations. The
maximum peak positions of Lyz–RxAc and Lyz–RxAcNPs were shifted slightly
toward a longer wavelength region (279–284 nm and 279–283 nm, respectively).
The change in λmax is observed possibly due to complex formation between Lyz
and RxAc/RxAcNPs. The red shift in the absorbance spectra also indicated that the
polarity of amino acid microenvironments increased with the addition of RxAc or
RxAcNPs [83–86], which is in good agreement with the quenching and Synchronous fluorescence spectroscopy and thermodynamic analysis results.

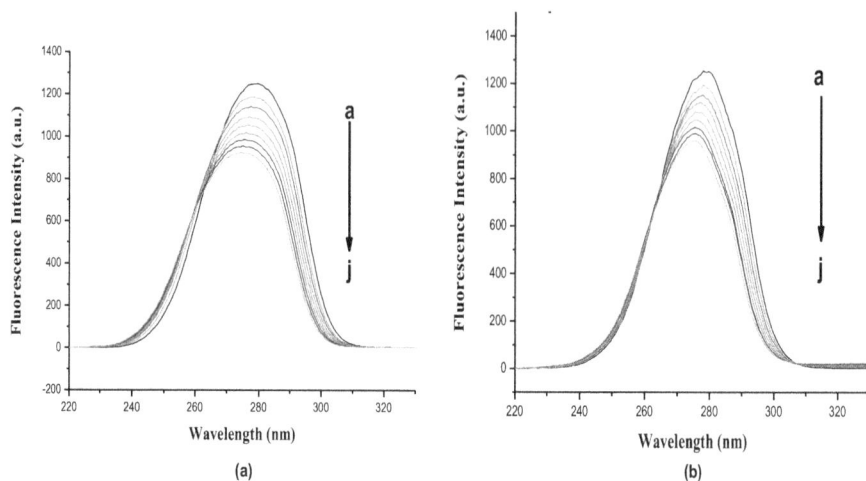

Figure 14.
Synchronous fluorescence spectrum of (A) Lyz-RxAc and (B) Lyz-RxAcNPs systems at 298 K : (Δλ = 60 nm), $C_{(Lyz)}$ = 10 μM; $C_{(RxAc\ or\ RxAcNPs)}$ (b-j): 2, 4, 6, 8, 10, 12, 14 and 16 μM.

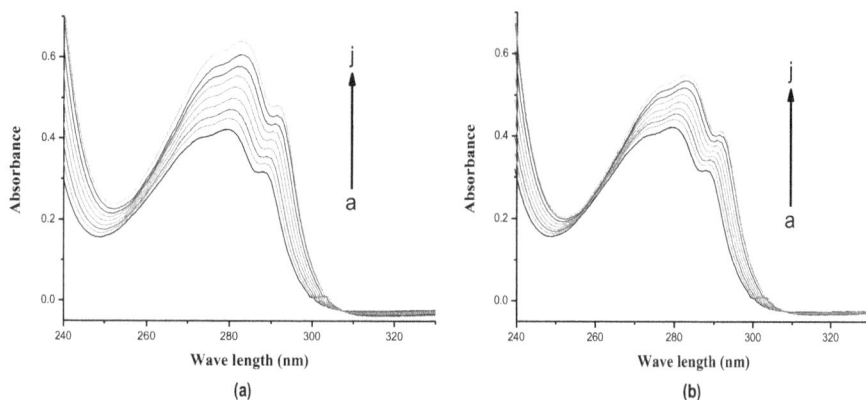

Figure 15.
UV–Vis spectra of Lyz in the presence of (A) RxAc and (B) RxAcNPs at 298 K. C_{Lyz}: 10 μM (a), $C_{RxAc\ or\ RxAcNPs}$ (b-j): 2, 4, 6, 8, 10, 12, 14 and 16 μM.

3.3.8 Circular dichroism spectroscopy

The technique of far-UV Circular dichroism spectroscopy (CD) is an important and powerful technology technique utilized to probe the secondary and tertiary structures of the protein/biopolymer [87–89]. The method is used to explore the biopolymer conformational changes upon binding of RxAc and RxAcNPs to Lyz, due to its simplicity and reliability. The CD spectra of Lyz with various concentrations of RxAc and RxAcNPs have been shown in **Figure 16A** and **B** at room temperature. The results of CD spectra of Lyz show two negative bands at 208 nm ($\pi \rightarrow \pi^*$ transition) and 229 nm (n $\rightarrow \pi^*$ transition), which are attributed to the α-helical structure of protein [67] whose magnitude reveals the amount of α-helicity in Lysozyme and they arise due to π–π^* and n–π^* transitions of the peptide bond of α-helix [84, 89–91]. The CD data have been observed in terms of mean residue ellipticity (MRE) in deg cm^{-2} dmol^{-1} according to the following Equation [92–97]:

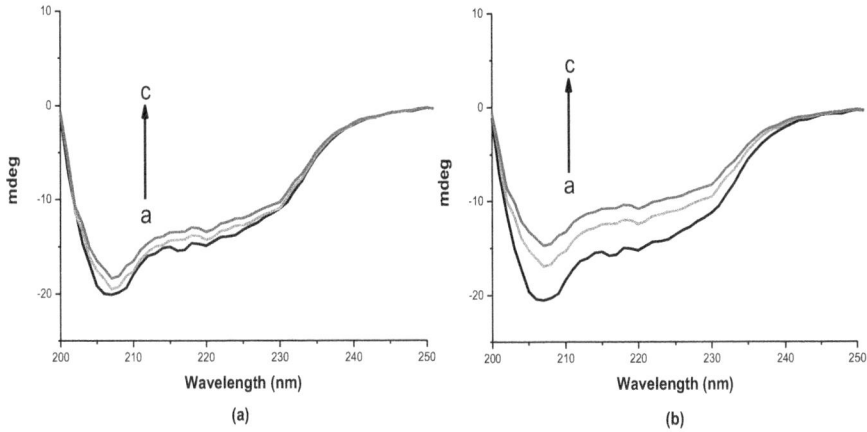

Figure 16.
The CD spectra of (A) Lyz-RxAc and (B) Lyz-RxAcNPs systems. Lyz concentration was kept fixed at 10 μM (a). In Lyz-RxAc and Lyz-RxAcNPs systems, RxAc or RxAcNPs concentration was fixed at 40 (b) and 80 μM (c).

$$MRE = \frac{obsCD(m \text{ deg})}{Cp \times n \times l \times 10} \tag{11}$$

where C_p is the molar concentration of protein, n is the number of amino acid residues of the protein (129 for Lyz), and l is the path length in cm. The α-helical content of Lyz is calculated from the MRE value at 208 nm, using the following Equation [92–97]:

$$\alpha - helix \ (\%) = \frac{-MRE_{208} - 4000}{33,000 - 4000} \times 100 \tag{12}$$

where MRE_{208} is the observed mean residue ellipticity (MRE) value at 208 nm, 4000 is the MRE of the β-form and random coil conformation cross at 208 nm, and 33,000 is the MRE value of a pure α-helix at 208 nm.

In order to study the influence of RxAc and RxAcNPs on the secondary structure of the Lyz, the CD measurements of Lyz in the absence and presence of RxAc and RxAcNPs were performed. From **Figure 16A,B** and **Table 7**, the α-helicity for free Lyz was 43.34%, while the addition of RxAc and RxAcNPs to the Lyz solution caused an increase in the negative peak ellipticities, probably as a consequence of the formation of complex between Lyz and RxAc/RxAcNPs. The CD data in the

System			α-helix %
[Lyz] (μM)	Drug	[Drug]	
		0	43.34
10	RxAc	40	41.12
		80	38.73
		0	43.34
10	RxAcNPs	40	35.82
		80	31.23

Table 7.
α-helicity (%) of Lyz at different concentrations of RxAc and RxAcNPs at 298 K.

wavelength range of 200–250 nm are used to evaluate the change of the secondary structure in Lyz. The results showed that interaction with RxAc and RxAcNPs caused only an increase in the band intensity of Lyz without any significant shift of the peaks and the helical content decreased to 38.73% and 31.23%, respectively. The decrease in the α-helical content of lysozyme represents the unfolding of protein due to interaction with RxAc or RxAcNPs. The unfolding of protein changes the absorbance value, which in turn alters the ellipticity value. The results showed that Lyz was induced to adopt a more loose conformation of the extended polypeptide. The conformational transition probably resulted in the exposure of the hydrophobic cavities to more hydrophilic environment, which is favorable for the interaction between Lyz and RxAc or RxAcNPs. The CD results also corroborate the conclusion of fluorescence and UV studies [77, 98].

3.3.9 Molecular docking analysis

The molecular docking is an excellent and an efficient computational technique, which is used to predict the probable binding site of interaction of protein with the ligand/drug molecule and also the preferred binding site of the ligand through the 3-D graphics [4, 11, 47, 99]. Moreover, it also displays the amino acid residues encircling the ligand/drug molecule and also assists to validate the results and highlights the interaction types operating in the protein–ligand/drug system [4, 99, 100]. To study the interaction in the Lyz–RxAc system, molecular docking studies were carried out with Autodock Vina software to clarify the mode of binding between lysozyme and Roxatidine acetate and illustrate the underlying mechanism. For further study, the lowest energy was chosen that was found to be −5.5 kcal/mol (−23.01 kj/mol), which is near to the experimental data of ΔG° in thermodynamic analysis (−22.25 kj/mol). In the Lyz–RxAc system (**Figure 17A** and **B**), the molecular docking showed the site of binding of RxAc along the long large pocket between the two domains of Lyz, which is also its site of active binding [101], and also showed that there were approximately three sites for binding of RxAc with lysozyme via hydrogen bonds at the lowest energy (Asn46 "two bonds" and Asp52 "one bond") (**Figure 18A** and **B**). The three sites of binding were all located in the large pocket of lysozyme. There were two hydrogen bonds between Asn46 and RxAc that were formed; one bond was between the hydrogen atom of Asn46 and the oxygen atom (C=O) of Roxatidine acetate, and another hydrogen bond was between the other hydrogen atom of Asn46 and the oxygen atom of Roxatidine

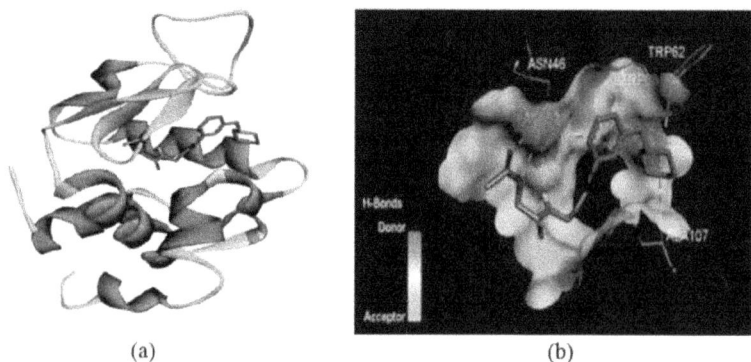

(a)   (b)

Figure 17.
Docking interaction of Lyz-RxAc binding, (A) The lysozyme pocket when RxAc was added. (B) Gaussian contact maps superimposed with the RxAc ligand and the receptor lysozyme, hydrogen bonds preference was indicated.

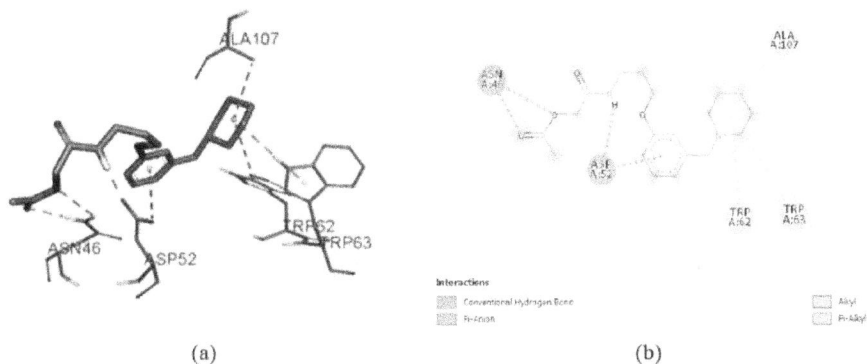

(a)   (b)

Figure 18.
(A) Amino acid residues surrounding RxAc. (B) 2-Dimentional representation of Lyz-RxAc system.

acetate (C—O—C). The third hydrogen bond was between the oxygen atom of Asp52 and the hydrogen atom (NH) of Roxatidine acetate. The dominant fluorophores (Trp62 and Trp63) were involved in the binding sites through hydrophobic interaction via Pi–alkyl binding, which could illustrate the observed quenching of fluorescence, whereas the aliphatic amino acid (Ala107) also formed hydrophobic interaction through alkyl–alkyl binding; in addition, the oxygen atom of Asp52 formed electrostatic binding with the phenyl ring of Roxatidine acetate, as shown in **Figures 17B** and **18A,B**. Molecular docking studies have explained that hydrogen bonding was the dominant driving force in the binding of RxAc to lysozyme, and these results were in accordance with the results of thermodynamic analysis and UV spectroscopy.

4. Conclusion

In the present study, the RxAc drug–loaded Tween80–Chitosan nanoparticles (RxAcNPs) have been characterized and probed through FTIR, PXRD, UV–Vis, DLS, and SEM techniques. The physicochemical properties of RxAcNPs have been employed and evaluated for drug formulation, determination of external morphology, particle size, drug content, entrapment efficiency, and in vitro release of drugs. In addition, the RxAc and RxAcNPs interactions with Lyz have been investigated utilizing spectroscopic methods such as fluorescence, UV–Vis, and CD spectroscopy. The results of Stern–Volmer plots illustrate that the interaction mechanism of Lyz–RxAc and Lyz–RxAcNPs systems was a static mechanism. In the presence of RxAc and RxAcNPs, the secondary construction of Lyz is reformed. The results of synchronous fluorescence and CD spectra confirm that the RxAc and RxAcNPs cause change in the secondary construction of Lyz. The thermodynamic results clarify that the main forces in both systems were hydrogen bonds and Van der Waals forces, also revealing that the reaction of binding in both systems is spontaneous, exothermic, and enthalpically driven. The molecular docking results were in accordance with the results of thermodynamic analysis, UV–Vis, and CD spectroscopy. The present study illustrates that the binding of RxAcNPs with Lyz is low as compared to RxAc, which confirms that the distribution and absorption of the RxAcNPs to various tissues would be higher. Therefore, the result of this study has a great importance in the clinical medicine and pharmacology area and provides important insight into the interaction of serum albumins with antiulcer drugs.

The significance of this study is evident because the developed RxAc-loaded Tween80–Chitosan nanoparticles could be utilized as an efficient strategy using nanotechnology in applications of ulcer therapy.

Acknowledgements

The authors thank the Department of Chemistry and Interdisciplinary Biotechnology Unit, Aligarh Muslim University, India, for support to this work. They also thank the authorities of Aligarh Muslim University for extending the necessary facilities.

Conflict of interest

The authors declare no conflict of interest.

Author details

Mohsen T.A. Qashqoosh[1,2*], Faiza A.M. Alahdal[1,3], Yahiya Kadaf Manea[1,2], Swaleha Zubair[4] and Saeeda Naqvi[1*]

1 Department of Chemistry, Aligarh Muslim University, Aligarh, Uttar Pradesh, India

2 Department of Chemistry, University of Aden, Aden, Yemen

3 Department of Chemistry, Hodeidah University, Hodeidah, Yemen

4 Department of Computer Science, Aligarh Muslim University, Aligarh, Uttar Pradesh, India

*Address all correspondence to: mohssenkashkush@yahoo.com and snaqvimo2015@gmail.com

IntechOpen

References

[1] Hegyi H, Gerstein M. The relationship between protein structure and function: A comprehensive survey with application to the yeast genome. Journal of Molecular Biology. 1999;**288**:147-164. DOI: 10.1006/JMBI.1999.2661

[2] Basdevant N, Weinstein H, Ceruso M. Thermodynamic basis for promiscuity and selectivity in protein-protein interactions: PDZ domains, a case study. Journal of the American Chemical Society. 2006;**128**:12766-12777. DOI: 10.1021/JA060830Y

[3] He HW, Zhang J, Zhou HM, Yan YB. Conformational change in the C-terminal domain is responsible for the initiation of creatine kinase thermal aggregation. Biophysical Journal. 2005; **89**:2650-2658. DOI: 10.1529/biophysj.105.066142

[4] Ansari SS, Yousuf I, Arjmand F, Siddiqi MK, Naqvi S. Exploring the intermolecular interactions and contrasting binding of flufenamic acid with hemoglobin and lysozyme: A biophysical and docking insight. International Journal of Biological Macromolecules. 2018;**116**:1105-1118. DOI: 10.1016/j.ijbiomac.2018.05.052

[5] Alam P, Siddiqi K, Chturvedi SK, Khan RH. Protein aggregation: From background to inhibition strategies. International Journal of Biological Macromolecules. 2017;**103**:208-219. DOI: 10.1016/j.ijbiomac.2017.05.048

[6] Chaturvedi SK, Siddiqi MK, Alam P, Khan RH. Protein misfolding and aggregation: Mechanism, factors and detection. Process Biochemistry. 2016; **51**:1183-1192. DOI: 10.1016/j.procbio.2016.05.015

[7] Saguer E, Alvarez P, Sedman J, Ramaswamy HS, Ismail AA. Heat-induced gel formation of plasma proteins: New insights by FTIR 2D correlation spectroscopy. Food Hydrocolloids. 2009;**23**:874-879. DOI: 10.1016/j.foodhyd.2008.03.013

[8] Maekawa H, Toniolo C, Broxterman QB, Ge NH. Two-dimensional infrared spectral signatures of 310- And α-helical peptides. The Journal of Physical Chemistry. B. 2007; **111**:3222-3235. DOI: 10.1021/JP0674874

[9] Ge F, Chen C, Liu D, Han B, Xiong X, Zhao S. Study on the interaction between theasinesin and human serum albumin by fluorescence spectroscopy. Journal of Luminescence. 2010;**130**: 168-173. DOI: 10.1016/j.jlumin.2009.08.003

[10] Wu LZ, Ma BL, Zou DW, Tie ZX, Wang J, Wang W. Influence of metal ions on folding pathway and conformational stability of bovine serum albumin. Journal of Molecular Structure. 2008;**877**:44-49. DOI: 10.1016/j.molstruc.2007.07.013

[11] Qashqoosh MTA, Alahdal FAM, Manea YK, Zakariya SM, Naqvi S. Synthesis, characterization and spectroscopic studies of surfactant loaded antiulcer drug into Chitosan nanoparticles for interaction with bovine serum albumin. Chemical Physics. 2019;**527**:110462. DOI: 10.1016/j.chemphys.2019.110462

[12] Qashqoosh MTA, Manea YK, Alahdal FAM, Naqvi S. Investigation of conformational changes of Bovine Serum Albumin upon binding with Benzocaine drug: A spectral and computational analysis. Bionanoscience. 2019;**9**:848-858. DOI: 10.1007/S12668-019-00663-7

[13] Vertegel AA, Siegel RW, Dordick JS. Silica nanoparticle size influences the structure and enzymatic activity of adsorbed lysozyme. Langmuir. 2004;**20**: 6800-6807. DOI: 10.1021/LA0497200

[14] Lee-Huang S, Maiorov V, Huang PL, Ng A, Hee CL, Chang YT, et al. Structural and functional modeling of human lysozyme reveals a unique nonapeptide, HL9, with anti-HIV activity. Biochemistry. 2005;**44**: 4648-4655. DOI: 10.1021/BI0477081

[15] Das A, Thakur R, Dagar A, Chakraborty A. A spectroscopic investigation and molecular docking study on the interaction of hen egg white lysozyme with liposomes of saturated and unsaturated phosphocholines probed by an anticancer drug ellipticine. Physical Chemistry Chemical Physics. 2014;**16**: 5368-5381. DOI: 10.1039/c3cp54247e

[16] Zhang HM, Chen J, Zhou QH, Shi YQ, Wang YQ. Study on the interaction between cinnamic acid and lysozyme. Journal of Molecular Structure. 2011;**987**:7-12. DOI: 10.1016/j. molstruc.2010.11.053

[17] Li D, Zhang T, Xu C, Ji B. Effect of pH on the interaction of vitamin B12 with bovine serum albumin by spectroscopic approaches. Spectrochimica Acta Part A: Molecular and Biomolecular Spectroscopy. 2011; **83**:598-608. DOI: 10.1016/j. saa.2011.09.012

[18] Banerjee S, Dutta Choudhury S, Dasgupta S, Basu S. Photoinduced electron transfer between hen egg white lysozyme and anticancer drug menadione. Journal of Luminescence. 2008;**128**:437-444. DOI: 10.1016/j. jlumin.2007.09.020

[19] Wang W, Min W, Chen J, Wu X, Hu Z. Binding study of diprophylline with lysozyme by spectroscopic methods. Journal of Luminescence. 2011;**131**:820-824. DOI: 10.1016/j. jlumin.2010.12.010

[20] Ibrahim HR, Matsuzaki T, Aoki T. Genetic evidence that antibacterial activity of lysozyme is independent of

its catalytic function. FEBS Letters. 2001;**506**:27-32. DOI: 10.1016/ S0014-5793(01)02872-1

[21] Croguennec T, Nau F, Molle D, Le Graet Y, Brule G. Iron and citrate interactions with hen egg white lysozyme. Food Chemistry. 2000;**68**: 29-35. DOI: 10.1016/S0308-8146(99) 00147-8

[22] Ghosh A, Brinda KV, Vishveshwara S. Dynamics of lysozyme structure network: Probing the process of unfolding. Biophysical Journal. 2007; **92**:2523-2535. DOI: 10.1529/ biophysj.106.099903

[23] Zhang Z, Zheng Q, Han J, Gao G, Liu J, Gong T, et al. The targeting of 14-succinate triptolide-lysozyme conjugate to proximal renal tubular epithelial cells. Biomaterials. 2009;**30**: 1372-1381. DOI: 10.1016/j. biomaterials.2008.11.035

[24] Gu Z, Zhu X, Ni S, Su Z, Zhou HM. Conformational changes of lysozyme refolding intermediates and implications for aggregation and renaturation. The International Journal of Biochemistry & Cell Biology. 2004; **36**:795-805. DOI: 10.1016/j. biocel.2003.08.015

[25] Murdoch D, McTavish D. Roxatidine Acetate. Drugs. 1991;**42**: 240-260. DOI: 10.2165/ 00003495-199142020-00006

[26] Walt RP, Logan RFA, Hawkey CJ, Daneshmend TK, Long RG, Cooper BT, et al. A comparison of roxatidine and ranitidine for the acute treatment of duodenal ulcer. Alimentary Pharmacology & Therapeutics. 1991;**5**: 301-307. DOI: 10.1111/J.1365-2036.1991. TB00031.X

[27] Bickel M, Herling AW, Schoelkens B, Scholtholt J. Chemical and biologic characteristics of roxatidine acetate. Scandinavian Journal of

Gastroenterology. 1988;**23**:78-88. DOI: 10.3109/00365528809099134

[28] Cataldo MG, Brancato D, Donatelli M. Comparison of roxatidine and ranitidine in the treatment of refractory duodenal ulcer. Current Therapeutic Research. 1994;**55**:438-445. DOI: 10.1016/S0011-393X(05)80530-2

[29] Guo X, Li X, Jiang Y, Yi L, Wu Q, Chang H, et al. A spectroscopic study on the interaction between p-nitrophenol and bovine serum albumin. Journal of Luminescence. 2014;**149**:353-360. DOI: 10.1016/j.jlumin.2014.01.036

[30] Desfougères Y, Saint-Jalmes A, Salonen A, Vié V, Beaufils S, Pezennec S, et al. Strong improvement of interfacial properties can result from slight structural modifications of proteins: The case of native and dry-heated lysozyme. Langmuir. 2011;**27**: 14947-14957. DOI: 10.1021/la203485y

[31] Dustgania A, Farahania EV, Imanib M. Preparation of Chitosan nanoparticles loaded by Dexamethasone Sodium Phosphate. Iranian Journal of Pharmaceutical Sciences. 2008;**4**: 111-114

[32] Agnihotri SA, Mallikarjuna NN, Aminabhavi TM. Recent advances on chitosan-based micro- and nanoparticles in drug delivery. Journal of Controlled Release. 2004;**100**:5-28. DOI: 10.1016/j.jconrel.2004.08.010

[33] Sri KV, Santhoshini G, Sankar DR, Niharika K. Formulation and evaluation of rutin loaded nanosponges. Asian Journal of Research in Pharmaceutical Sciences. 2018;**8**:21. DOI: 10.5958/2231-5659.2018.00005.x

[34] Jeevitha D, Amarnath K. Chitosan/PLA nanoparticles as a novel carrier for the delivery of anthraquinone: Synthesis, characterization and in vitro cytotoxicity evaluation. Colloids and Surfaces. B, Biointerfaces. 2013;**101**:

126-134. DOI: 10.1016/j.colsurfb.2012.06.019

[35] Vora C, Patadia R, Mittal K, Mashru R. Formulation development, process optimization, and in vitro characterization of spray-dried lansoprazole enteric microparticles. Scientia Pharmaceutica. 2016;**84**: 393-408. DOI: 10.3797/scipharm.1501-08

[36] Ruckmani K, Sivakumar M, Satheesh Kumar S. Nanoparticular drug delivery system of cytarabine hydrochloride (CTH) for improved treatment of lymphoma, in. Journal of Biomedical Nanotechnology. 2007;**3**(1): 90-96. DOI: 10.1166/jbn.2007.016

[37] Das S, Banerjee R, Bellare J. Aspirin loaded albumin nanoparticles by coacervation: Implications in drug delivery. Biomaterials and Artificial Organs. 2005;**18**(2):203-212

[38] Nagarajan E, Shanmugasundaram P, Ravichandiran V, Vijayalakshmi A, Senthilnathan B, Masilamani K. Development and evaluation of chitosan based polymeric nanoparticles of an antiulcer drug Lansoprazole. Journal of Applied Pharmaceutical Science. 2015; **5**:20-25. DOI: 10.7324/JAPS.2015.50404

[39] Elshafeey AH, Kamel AO, Awad GAS. Ammonium methacrylate units polymer content and their effect on acyclovir colloidal nanoparticles properties and bioavailability in human volunteers. Colloids and Surfaces. B, Biointerfaces. 2010;**75**:398-404. DOI: 10.1016/j.colsurfb.2009.08.050

[40] Anand M, Sathyapriya P, Maruthupandy M, Hameedha Beevi A. Synthesis of chitosan nanoparticles by TPP and their potential mosquito larvicidal application. Frontiers in Laboratory Medicine. 2018;**2**:72-78. DOI: 10.1016/j.flm.2018.07.003

[41] Anand M, Maruthupandy M, Kalaivani R, Suresh S, Kumaraguru AK. Larvicidal activity of Chitosan Nanoparticles Synthesized from Crab and Squilla Species against Aedes aegypti. Journal of Colloid Science and Biotechnology. 2015;**3**:188-193. DOI: 10.1166/jcsb.2014.1088

[42] Song R, Xue R, He LH, Liu Y, Xiao QL. The structure and properties of chitosan/polyethylene glycol/silica ternary hybrid organic-inorganic films. Chinese Journal of Polymer Science (English Edition). 2008;**26**:621-630. DOI: 10.1142/S0256767908003357

[43] George M, Abraham TE. Polyionic hydrocolloids for the intestinal delivery of protein drugs: Alginate and chitosan - a review. Journal of Controlled Release. 2006;**114**:1-14. DOI: 10.1016/j.jconrel. 2006.04.017

[44] Namasivayam S, Robin A. Preparation of nano albumin-flutamide (Nab-flu) conjugate and evaluation of its in vitro drug control release, anticancer activity and genotoxicity. Indian Journal of Experimental Biology. 2018;**56**:171-179

[45] Raj V, Prabha G. Synthesis, characterization and in vitro drug release of cisplatin loaded Cassava starch acetate–PEG/gelatin nanocomposites. Journal of the Association of Arab Universities for Basic and Applied Sciences. 2016;**21**: 10-16. DOI: 10.1016/j. jaubas.2015.08.001

[46] Imoto T, Forster LS, Rupley JA, Tanaka F. Fluorescence of lysozyme: Emissions from tryptophan residues 62 and 108 and energy migration. Proceedings of the National Academy of Sciences of the United States of America. 1972;**69**:1151-1155. DOI: 10.1073/pnas.69.5.1151

[47] Jing M, Song W, Liu R. Binding of copper to lysozyme: Spectroscopic, isothermal titration calorimetry and

molecular docking studies. Spectrochimica Acta Part A: Molecular and Biomolecular Spectroscopy. 2016; **164**:103-109. DOI: 10.1016/j. saa.2016.04.008

[48] Millan S, Satish L, Kesh S, Chaudhary YS, Sahoo H. Interaction of Lysozyme with Rhodamine B: A combined analysis of spectroscopic & molecular docking. Journal of Photochemistry and Photobiology B: Biology. 2016;**162**:248-257. DOI: 10.1016/j.jphotobiol.2016.06.047

[49] Revathi R, Rameshkumar A, Sivasudha T. Spectroscopic investigations on the interactions of AgTiO2 nanoparticles with lysozyme and its influence on the binding of lysozyme with drug molecule. Spectrochimica Acta Part A: Molecular and Biomolecular Spectroscopy. 2016;**152**:192-198. DOI: 10.1016/j.saa.2015.07.066

[50] Siddiqi M, Nusrat S, Alam P, Malik S, Chaturvedi SK, Ajmal MR, et al. Investigating the site selective binding of busulfan to human serum albumin: Biophysical and molecular docking approaches. International Journal of Biological Macromolecules. 2018;**107**: 1414-1421. DOI: 10.1016/j. ijbiomac.2017.10.006

[51] Shen H, Gu Z, Jian K, Qi J. In vitro study on the binding of gemcitabine to bovine serum albumin. Journal of Pharmaceutical and Biomedical Analysis. 2013;**75**:86-93. DOI: 10.1016/j. jpba.2012.11.021

[52] Wang Z, Tan X, Chen D, Yue Q, Song Z. Study on the binding behavior of lysozyme with cephalosporin analogues by fluorescence spectroscopy. Journal of Fluorescence. 2009;**19**: 801-808. DOI: 10.1007/S10895-009-0477-8

[53] Lakowicz J. Principles of fluorescence spectroscopy. 2013. DOI: 10.1007/978-0-387-46312-4

[54] Roy S. Review on interaction of serum albumin with drug molecules research and reviews. Journal of Pharmacology and Toxicological Studies. 2016;**4**(2):7-16

[55] Wang G, Hou H, Wang S, Yan C, Liu Y. Exploring the interaction of silver nanoparticles with lysozyme: Binding behaviors and kinetics. Colloids and Surfaces. B, Biointerfaces. 2017;**157**: 138-145. DOI: 10.1016/j. colsurfb.2017.05.071

[56] Melavanki RM, Kusanur RA, Kadadevaramath JS, Kulakarni MV. Quenching mechanisms of 5BAMC by aniline in different solvents using Stern-Volmer plots. Journal of Luminescence. 2009;**129**:1298-1303. DOI: 10.1016/j. jlumin.2009.06.011

[57] Papadopoulou A, Green RJ, Frazier RA. Interaction of flavonoids with bovine serum albumin: A fluorescence quenching study. Journal of Agricultural and Food Chemistry. 2005;**53**:158-163. DOI: 10.1021/ JF048693G

[58] Evale BG, Hanagodimath SM. Fluorescence quenching of newly synthesized biologically active coumarin derivative by aniline in binary solvent mixtures. Journal of Luminescence. 2009;**129**:1174-1180. DOI: 10.1016/j. jlumin.2009.05.017

[59] Melavanki RM, Kusanur RA, Kulakarni MV, Kadadevarmath JS. Role of solvent polarity on the fluorescence quenching of newly synthesized 7,8-benzo-4-azidomethyl coumarin by aniline in benzene-acetonitrile mixtures. Journal of Luminescence. 2008;**128**: 573-577. DOI: 10.1016/j. jlumin.2007.08.013

[60] Zhang Q, Ni Y, Kokot S. Competitive interactions between glucose and lactose with BSA: Which sugar is better for children? Analyst. 2016;**141**:2218-2227. DOI: 10.1039/c5an02420j

[61] Sohrabi Y, Panahi-Azar V, Barzegar A, Dolatabadi JEN, Dehghan P. Spectroscopic, thermodynamic and molecular docking studies of bovine serum albumin interaction with ascorbyl palmitate food additive. BioImpacts: BI. 2017;**7**:241-246. DOI: 10.15171/bi.2017.28

[62] Li D, Yang Y, Cao X, Xu C, Ji B. Investigation on the pH-dependent binding of vitamin B12 and lysozyme by fluorescence and absorbance. Journal of Molecular Structure. 2012;**1007**:102-112. DOI: 10.1016/j.molstruc.2011.10.028

[63] Ercelen S, Klymchenko AS, Mély Y, Demchenko AP. The binding of novel two-color fluorescence probe FA to serum albumins of different species. International Journal of Biological Macromolecules. 2005;**35**:231-242. DOI: 10.1016/j.ijbiomac.2005.02.002

[64] Naveenraj S, Anandan S. Binding of serum albumins with bioactive substances - Nanoparticles to drugs. Journal of Photochemistry and Photobiology C Photochemistry Reviews. 2013;**14**:53-71. DOI: 10.1016/j. jphotochemrev.2012.09.001

[65] Zhang HX, Huang X, Zhang M. Thermodynamic studies on the interaction of dioxopromethazine to β-cyclodextrin and bovine serum albumin. Journal of Fluorescence. 2008;**18**: 753-760. DOI: 10.1007/S10895-008-0348-8

[66] Ao J, Gao L, Yuan T, Jiang G. Interaction mechanisms between organic UV filters and bovine serum albumin as determined by comprehensive spectroscopy exploration and molecular docking. Chemosphere. 2015;**119**:590-600. DOI: 10.1016/j.chemosphere.2014.07.019

[67] Kumar CV, Buranaprapuk A, Sze HC, Jockusch S, Turro NJ. Chiral protein scissors: High enantiomeric selectivity for binding and its effect on

protein photocleavage efficiency and specificity. Proceedings of the National Academy of Sciences of the United States of America. 2002;**99**:5810-5815. DOI: 10.1073/pnas.082119599

[68] Calleri E, De Lorenzi E, Siluk D, Markuszewski M, Kaliszan R, Massolini G. Riboflavin binding protein - chiral stationary phase: Investigation of retention mechanism. Chromatographia. 2002;**55**:651-658. DOI: 10.1007/BF02491778

[69] Fathi F, Mohammadzadeh-Aghdash H, Sohrabi Y, Dehghan P, Dolatabadi JEN. Kinetic and thermodynamic studies of bovine serum albumin interaction with ascorbyl palmitate and ascorbyl stearate food additives using surface plasmon resonance. Food Chemistry. 2018;**246**: 228-232. DOI: 10.1016/j.foodchem.2017.11.023

[70] Chirio-Lebrun MC, Prats M. Fluorescence resonance energy transfer (FRET): Theory and experiments. Biochemical Education. 1998;**26**:320-323. DOI: 10.1016/S0307-4412(98)80010-1

[71] Naik KM, Nandibewoor ST. Spectroscopic studies on the interaction between chalcone and bovine serum albumin. Journal of Luminescence. 2013;**143**:484-491. DOI: 10.1016/j.jlumin.2013.05.013

[72] Howell WM, Jobs M, Brookes AJ. iFRET: An improved fluorescence system for DNA-melting analysis. Genome Research. 2002;**12**:1401-1407. DOI: 10.1101/gr.297202

[73] Ilichev YV, Perry JL, Simon JD. Interaction of ochratoxin a with human serum albumin. Preferential binding of the dianion and pH effects. The Journal of Physical Chemistry. B. 2002;**106**: 452-459. DOI: 10.1021/JP012314U

[74] Zhou N, Liang YZ, Wang P. 18β-Glycyrrhetinic acid interaction with

bovine serum albumin. Journal of Photochemistry and Photobiology A: Chemistry. 2007;**185**:271-276. DOI: 10.1016/j.jphotochem.2006.06.019

[75] Roy S. Binding behaviors of greenly synthesized silver nanoparticles – Lysozyme interaction: Spectroscopic approach. Journal of Molecular Structure. 2018;**1154**:145-151. DOI: 10.1016/j.molstruc.2017.10.048

[76] Roy S. Binding behaviors of greenly synthesized silver nanoparticles – Lysozyme interaction: Spectroscopic approach. Journal of Molecular Structure. 2018;**1154**:145-151. DOI: 10.1016/j.molstruc.2017.10.048

[77] Ajmal MR, Abdelhameed AS, Alam P, Khan RH. Interaction of new kinase inhibitors cabozantinib and tofacitinib with human serum alpha-1 acid glycoprotein. A comprehensive spectroscopic and molecular Docking approach. Spectrochimica Acta Part A: Molecular and Biomolecular Spectroscopy. 2016;**159**:199-208. DOI: 10.1016/j.saa.2016.01.049

[78] Wu YL, He F, He XW, Li WY, Zhang YK. Spectroscopic studies on the interaction between CdTe nanoparticles and lysozyme, Spectrochim. Spectrochimica Acta, Part A: Molecular and Biomolecular Spectroscopy. 2008; **71**:1199-1203. DOI: 10.1016/j.saa.2008.03.018

[79] Manea YK, Khan AMT, Qashqoosh MTA, Wani AA, Shahadat M. Ciprofloxacin-supported chitosan/polyphosphate nanocomposite to bind bovine serum albumin: Its application in drug delivery. Journal of Molecular Liquids. 2019;**292**. DOI: 10.1016/j.molliq.2019.111337

[80] Ashoka S, Seetharamappa J, Kandagal PB, Shaikh SMT. Investigation of the interaction between trazodone hydrochloride and bovine serum albumin. Journal of Luminescence.

2006;**121**:179-186. DOI: 10.1016/j.jlumin.2005.12.001

[81] Paramaguru G, Kathiravan A, Selvaraj S, Venuvanalingam P, Renganathan R. Interaction of anthraquinone dyes with lysozyme: Evidences from spectroscopic and docking studies. Journal of Hazardous Materials. 2010;**175**:985-991. DOI: 10.1016/j.jhazmat.2009.10.107

[82] Bi S, Song D, Tian Y, Zhou X, Liu Z, Zhang H. Molecular spectroscopic study on the interaction of tetracyclines with serum albumins. Spectrochimica Acta Part A: Molecular and Biomolecular Spectroscopy. 2005;**61**:629-636. DOI: 10.1016/j.saa.2004.05.028

[83] Shaikh SMT, Seetharamappa J, Kandagal PB, Ashoka S. Binding of the bioactive component isothipendyl hydrochloride with bovine serum albumin. Journal of Molecular Structure. 2006;**786**:46-52. DOI: 10.1016/j.molstruc.2005.10.021

[84] Hemalatha K, Madhumitha G, Al-Dhabi NA, Arasu MV. Importance of fluorine in 2,3-dihydroquinazolinone and its interaction study with lysozyme. Journal of Photochemistry and Photobiology B: Biology. 2016;**162**:176-188. DOI: 10.1016/j.jphotobiol.2016.06.036

[85] Liu X, Li H. Spectroscopic studies on the interaction of polydatin with bovine serum albumin. Asian Journal of Chemistry. 2013;**25**:8131-8135. DOI: 10.14233/ajchem.2013.15264

[86] Jash C, Basu P, Payghan PV, Ghoshal N, Kumar GS. Chelerythrine-lysozyme interaction: Spectroscopic studies, thermodynamics and molecular modeling exploration. Physical Chemistry Chemical Physics. 2015;**17**:16630-16645. DOI: 10.1039/c5cp00424a

[87] Corrêa D, Ramos C. The use of circular dichroism spectroscopy to study protein folding, form and function. African Journal of Biochemistry Research. 2009;**3**:164-173. DOI: 10.5897/AJBR.9000245

[88] Chaturvedi SK, Khan JM, Siddiqi MK, Alam P, Khan RH. Comparative insight into surfactants mediated amyloidogenesis of lysozyme. International Journal of Biological Macromolecules. 2016;**83**:315-325. DOI: 10.1016/j.ijbiomac.2015.11.053

[89] Ding F, Zhao G, Huang J, Sun Y, Zhang L. Fluorescence spectroscopic investigation of the interaction between chloramphenicol and lysozyme. European Journal of Medicinal Chemistry. 2009;**44**:4083-4089. DOI: 10.1016/j.ejmech.2009.04.047

[90] Rahman MH, Maruyama T, Okada T, Yamasaki K, Otagiri M. Study of interaction of carprofen and its enantiomers with human serum albumin-I. Mechanism of binding studied by dialysis and spectroscopic methods. Biochemical Pharmacology. 1993;**46**:1721-1731. DOI: 10.1016/0006-2952(93)90576-I

[91] Azimi O, Emami Z, Salari H, Chamani J. Probing the interaction of human serum albumin with norfloxacin in the presence of high-frequency electromagnetic fields: Fluorescence spectroscopy and circular dichroism investigations. Molecules. 2011;**16**:9792-9818. DOI: 10.3390/molecules16129792

[92] Greenfield NJ. Using circular dichroism spectra to estimate protein secondary structure. Nature Protocols. 2007;**1**:2876-2890. DOI: 10.1038/nprot.2006.202

[93] Kelly S, Price N. The use of circular dichroism in the investigation of protein structure and function. Current Protein & Peptide Science. 2005;**1**:349-384. DOI: 10.2174/1389203003381315

[94] Ranjbar B, Gill P. Circular dichroism techniques: Biomolecular and nanostructural analyses- A review. Chemical Biology & Drug Design. 2009;**74**:101-120. DOI: 10.1111/j.1747-0285.2009.00847.x

[95] Price NC. Conformational issues in the characterization of proteins. Biotechnology and Applied Biochemistry. 2000;**31**:29. DOI: 10.1042/BA19990102

[96] Dolatabadi JEN, Panahi-Azar V, Barzegar A, Jamali AA, Kheirdoosh F, Kashanian S, et al. Spectroscopic and molecular modeling studies of human serum albumin interaction with propyl gallate. RSC Advances. 2014;**4**: 64559-64564. DOI: 10.1039/c4ra11103f

[97] Bhogale A, Patel N, Mariam J, Dongre PM, Miotello A, Kothari DC. Comprehensive studies on the interaction of copper nanoparticles with bovine serum albumin using various spectroscopies. Colloids and Surfaces. B, Biointerfaces. 2014;**113**:276-284. DOI: 10.1016/j.colsurfb.2013.09.021

[98] Parker W, Song PS. Protein structures in SDS micelle-protein complexes. Biophysical Journal. 1992;**61**: 1435-1439. DOI: 10.1016/S0006-3495 (92)81949-5

[99] Hernández-Santoyo A, Tenorio-Barajas AY, Altuzar V, Vivanco-Cid H, Mendoza-Barrera C. Protein engineering-technology and application. IntechOpen. 2013:63-81. DOI: 10.5772/56376

[100] Leelananda SP, Lindert S. Computational methods in drug discovery. Beilstein Journal of Organic Chemistry. 2016;**12**:2694-2718. DOI: 10.3762/bjoc.12.267

[101] Callewaert L, Michiels CW. Lysozymes in the animal kingdom. Journal of Biosciences. 2010;**35**:127-160. DOI: 10.1007/s12038-010-0015-5

Action of Surfactants in Driving Ecotoxicity of Microplastic-Nano Metal Oxides Mixtures: A Case Study on *Daphnia magna* under Different Nutritional Conditions

Cristiana Guerranti, Serena Anselmi, Francesca Provenza, Andrea Blašković and Monia Renzi

Abstract

The series of experiments presented in the paper served to clarify the effects of contemporary exposure to surfactant, microplastics (polyethylene and polyvinyl chloride), and nanoparticles (TiO_2 and ZnO) on the model organism *Daphnia magna*. Exposure was evaluated with respect to the age of the organisms ("young", 24 hours old, and "aged" 10 days old specimens), trophic status (feeding or fasting), and the simultaneous presence of a surfactant. All the above-mentioned substances are present in the wastewater coming from various environmental sources from cosmetic products. The experiments were conducted in compliance with the OECD 202:2004 guideline, which is also a reference for ecotoxicity tests required by REACH. The results showed that surfactants enhance effects of toxicity produced by the exposure to the microplastic + nanoparticle mixtures. The influence due to factors such as nutrition (effect in fasting >> feeding conditions) and the age of individuals (effects in older >> younger animals) is essential. Concerning young individuals, exposure to PE-TiO_2 is the most significant in terms of effects produced: it is very significant, especially in the presence of surfactant (both under fasting and feeding conditions). On the contrary, exposure to the PE-Zn mixture shows the minor effects. The comparison with the literature, especially as regards the possibility of interpreting the toxicity trends for the various mixtures with respect to the individual elements that compose them, leads to hypothesize additive effects still to be investigated and confirms the greatest toxicity contribution of TiO_2.

Keywords: metal-oxide nanoparticles, surfactants, fasting and feeding conditions, microplastics, municipal wastewater treatment plants, toxicity tests

1. Introduction

Municipal wastewater treatment plants (MWWTP) are renowned hot-spot sources of a wide variety of pollutants from human activities for the aquatic environments [1]. Nutrients [1], surfactants [2], microplastics (MPs; [3]) and nanoparticles

(NPs; [4]) are significant components of mixtures released by MWWTP, coming from different sources, not least that represented by cosmetic products.

NPs are chemicals, between 1 and 100 nm in size [5], both of natural (i.e., humic and fulvic acids, organic acids, fullerenes, and metals) and artificial origin (TiO_2, ZnO) [6, 7]. Many products of common use (pharmaceutical and personal care products, plastic, rubber, paints, etc) are based on NPs and this contributes to their massive presence in the environment [8].

MPs can derive from industrial pellets used in the manufacture of plastic objects or from consumer products, such as cosmetics, abrasive products and objects containing microbeads and glitters (primary microplastics) or from the fragmentation of larger plastic objects (secondary microplastics) [3, 9–11]. They are pollutant of great environmental concern [12], affecting feeding habits, and reproductive success of many organisms [13]. NPs and MPs enter the trophic chains when eaten by detritivores and filter feeders [14–17].

Even if wastewater treatment processes retain a large fraction of plastic microparticles [18], in sewage the MPs not removed by plants reach the rivers and, ultimately, the sea [19]. For what concerns NPs, discharges of nano-oxides occur due the sorptive removal of organic contaminants from wastewater [20] purification purposes, together with unintentional releases.

Both, MPs, and NPs are found in the mix of substances present in wastewater and water bodies, together with surfactants. Surfactants have direct toxicity on aquatic species [21] but can also vehicle other substances due to the formation of micelles [22] which can affect pollutant sorption/desorption from MPs surfaces [23]. Surfactants could represent the way to increase the interaction among microplastics and animals and therefore lead to negative effects on exposed animals [22].

Toxic effects due to the exposure to complex mixtures differ from the exposure to single substances, even if compounds are at low levels [24]. In this case, NPs and MPs toxicity could be affected both by the nutrient induced microalgal growth and by synergic/antagonistic interactive effects due to surfactants [25]. The presence of metal-oxides NPs, MPs, and surfactants in effluents from MWWTP suggests deepening their ecotoxicity to assess the real effects on aquatic environments.

Despite the increasing interest, recent meta-data analysis underlined low standardization of in vitro tests [25] and the largest number of experiments performed on single kind MP/NP. In Europe, the placing on the market of new formulations implies the verification of their environmental compatibility in accordance with the current regulation (REACH).

Daphnia magna (Cladocera) is a crucial model freshwater species for ecotoxicological tests due to the well standardized procedural practice (i.e., OECD standards) for experimental exposure [26].

This study aims to fill some important knowledge gaps on NPs and MPs ecotoxicity. It evaluates ecotoxicological responses of *D. magna* exposed complex mixtures (NPs + MPs in presence/absence of surfactant). Effects attained under fasting conditions (OECD guidelines) are matched to those achieved following the exposure of animals under feeding conditions, supposed as natural and with animals of different ages.

2. Material and methods

2.1 Experimental design

The experiments were performed under the OECD 202:2004 guideline [27]. Dispersions were made by suspension of tested particles in UNI EN ISO 6341:2012 standard freshwater. Experiments previously reported [28, 29] allowed to determine for each toxicant the dose permitting the survival of a significant fraction of the tested population until the end of the exposure time (96 h).

Figure 1 summarizes the experimental design. Mixtures of NPs (n-ZnO and n-TiO$_2$) and MPs (PE and PVC) were used for the exposure experiments, adding/not adding a non-ionic surfactant (Triton X-100, CAS n. 9002-93-1; tested at 0.001% v/v according to [22]) to improve the dispersion of MPs and NPs in tested samples; results were compared to controls to test ecotoxicological effects of the NPs-MPs and NPs-MPs-surfactant mixtures. Furthermore, we also exposed to dispersions of microplastics + surfactant animals that at the beginning of the experiment had 10 days of life (called "aged") to evaluate the effect of aging on the ecotoxicological responses. Animals were exposed under fasting conditions, selecting immobilization as endpoint and a contact time 24-48 h. Contextually to standard conditions required by OECD, animals were also exposed under feeding conditions and contact time was extended from 24 to 48 h to 96 h as suggested by the literature for tests performed on particulate toxicant [30] performing observations daily starting from T$_{24}$ after the initial exposure. Experiments were made during an 8:16 dark/light exposure cycle [31].

ZnO and n-TiO$_2$ were tested at 1.12 and 113.18 mg/L respectively; microplastic doses were 0.05 mg/L. The selection of microplastics to be tested was done according to Renzi et al. [29].

Dispersions of both single metal-oxides NPs, MPs and mixtures were characterized by microscopy coupled to Fourier Transformed-Infrared spectrometer (μFT-IR, model Nicolet i-10 MX equipped with ATR detector, Thermo®). The formation of clusters of nanoparticles in the mixtures was verified at the micrometric scale even in conditions with the addition of food. Survival rates % of exposed animals compared to negative controls were used as target endpoints.

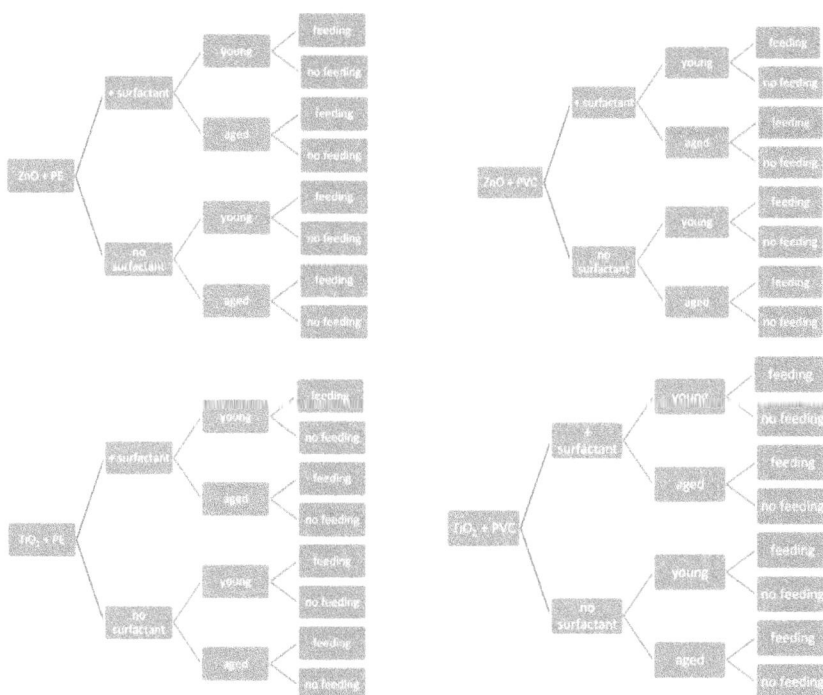

Figure 1.
Logic model of the exposure experiments on D. magna.

2.2 Equipment and materials

Experimental condition for *D. magna* (MicroBioTest Inc. ephippia) storage, hatches and preliminary treatments were the same described by Renzi and Blašković [28]. Collection of organisms was standardized at 90 h after the start of incubation. All reagents were purchased from Caelo or Sigma-Aldrich. Experimental conditions and sets were the same previously described by Renzi et al. [29] and Renzi and Blašković [28] to allow a comparison of reported data on toxicity of mixtures tested in this study with results obtained by previous research on single components of the tested mixtures. Also, experimental conditions are the same to allow a complete comparability of obtained results. Chemicals features and properties of nanoparticles, microplastic, and surfactant tested in this study are the same for chemicals reported by Renzi et al. [29] and Renzi and Blašković [28].

2.3 Quality assurance and quality control

Ecotoxicological tests were performed following UNI EN ISO 17025 guidelines to ensure quality control of collected results. Laboratory performed experiments ensured to pass inter-calibration exercises performed on annual basis on *D. magna* immobilization. During experiments pH and DOM were measured to verify that they remained within the limits of acceptability defined by OECD (202:2004) [27]. Mortality of animals in negative controls shall be included within 0-10% not to invalidate tests. LC_{50} of chemical solution used as reference to test animals' responses was acceptable (0.6–2.1 mg/L). Test were performed in triplicates (n=3); results obtained by the exposure to samples were compared statistically with responses obtained by negative controls; significance (p<0.01) of observed differences among mean values was tested by T-test while differences among variances were explored by F-test (Prism® 4.0). Results reported in this study are mean values (standard deviation, SD), normalized concerning negative controls.

3. Results and discussion

The results obtained from the exposure experiments are summarized in **Table 1** (young specimens) and **Table 2** (aged specimens). In the following text, the term significant refers to statistically significant based on the results of the statistical analysis carried out on the results obtained from the exposure experiments considering immobilization as the endpoint.

As for both young and aged individuals, exposure to PE-TiO_2 is the most significant in terms of effects: it is very significant, especially in the presence of triton (both fasting and feeding). On the contrary, exposure to the PE-Zn mixture shows the minor effects (**Figure 2**). This result, not linear if we consider that there is no clearly identifiable trend in the toxicity levels of the NPs/MPs, leads to think that more than the actual toxicity of the individual elements of the mixtures, there is a significant additivity effect. This estimation would also be legitimized by the comparison with previously obtained results: PVC, in presence of surfactant, resulted the most toxic among tested dispersions for both neonates and aged *Daphnia magna* exposed [29].

On the other hand, in Renzi and Blašković [28], n-ZnO resulted less effective than n-TiO_2 in leading to the target endpoints (immobilization and death) in exposed *Daphnia magna* individuals; in this sense the observations made in this study would confirm a higher toxicity contribution of n-TiO_2 compared to n-ZnO. A further factor to be taken into consideration in the interpretation and future

		Feeding conditions (MPs 0.05 mg/L; n-TiO$_2$ 113.18 mg/L)				Feeding conditions (MPs 0.05 mg/L; n-ZnO 1.12 mg/L)			
	Surfactant	no	no	yes	yes	no	no	yes	yes
	MPs type	PE	PVC	PE	PVC	PE	PVC	PE	PVC
24 h	%Effect	6.67	6.67	17.78	6.67	0.00	0.00	13.33	6.67
	DS	0.00	0.00	9.62	11.55	0.00	0.00	0.00	11.55
96 h	%Effect	20.00	26.67	46.67	60.00	6.67	33.33	26.67	73.33
	DS	0.00	11.55	11.55	11.55	0.00	11.55	11.55	0.00
		Fasting conditions (MPs 0.05 mg/L; n-TiO$_2$ 113.18 mg/L)				Fasting conditions (MPs 0.05 mg/L; n-ZnO 1.12 mg/L)			
	Surfactant	no	no	yes	yes	no	no	yes	yes
	MPs type	PE	PVC	PE	PVC	PE	PVC	PE	PVC
24 h	%Effect	6.67	13.33	46.67	46.67	0.00	86.67	0.00	33.33
	DS	11.55	0.00	20.00	0.00	0.00	11.55	0.00	20.00
96 h	%Effect	53.33	60.00	100.00	93.33	46.67	100.00	80.00	100.00
	DS	11.55	0.00	11.55	11.55	11.55	11.55	0.00	11.55

Surfactant (Triton X-100).

Table 1.
Percentages observed for the "immobilization" endpoint in young individuals, in relation to experimental exposure parameters.

		Feeding conditions (MPs 0.05 mg/L; n-TiO$_2$ 113.18 mg/L; Surfactant)		Feeding conditions (MPs 0.05 mg/L; n-ZnO 1.12 mg/L; Surfactant)	
	MPs Type	PE	PVC	PE	PVC
24 h	%Effect	20.00	6.67	6.67	0.00
	DS	11.55	11.55	11.55	0.00
96 h	%Effect	93.33	93.33	53.33	46.67
	DS	0.00	11.55	11.55	11.55
		Fasting conditions (MPs 0.05 mg/L; n-TiO$_2$ 113.18 mg/L; Surfactant)		Fasting conditions (MPs 0.05 mg/L; n ZnO 1.12 mg/L; Surfactant)	
	MPs Type	PE	PVC	PE	PVC
24 h	%Effect	40.00	0.00	20.00	6.67
	DS	20.00	0.00	23.09	11.55
96 h	%Effect	93.33	80.00	100.00	100.00
	DS	11.55	20.00	0.00	0.00

Surfactant (Triton X-100).

Table 2.
Percentages observed for the "immobilization" endpoint in aged individuals, in relation to experimental exposure parameters.

analysis of the results is the ability to interact at a chemical level between surfactant and MPs, which could be material dependent. This difference between plastics of different nature, already hypothesized [32], could explain the variability of the results obtained in terms of toxicity of different mixtures.

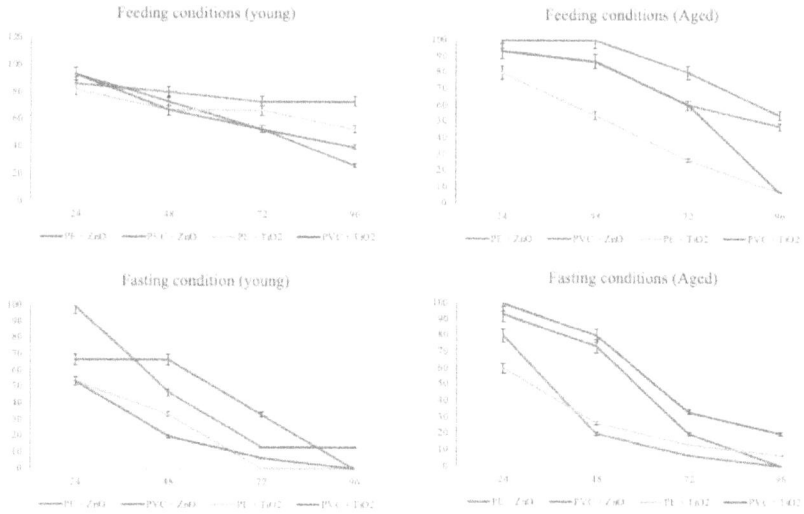

Figure 2.
Comparison between young and adult organisms exposed in surfactant presence and fasting/feeding. Graphics report the mean percentages of mobile organisms at the target time normalized for negative controls.

3.1 Surfactant effect

The effect of surfactant (Triton X-100) was significant in all the experiments carried out on young organisms, compared to the control batches without the exposure to surfactant. For this reason, and to simplify the factors considered, making the observed effects more evident, it was decided to always add surfactant to the mixtures of contaminants to which adult organisms were exposed.

Figure 3 shows the contribution of surfactant presence/absence under different trophic conditions in young organisms exposed to NPs+MPs mixtures, in terms of percentages of mobile organisms at the target times. These effects could be because

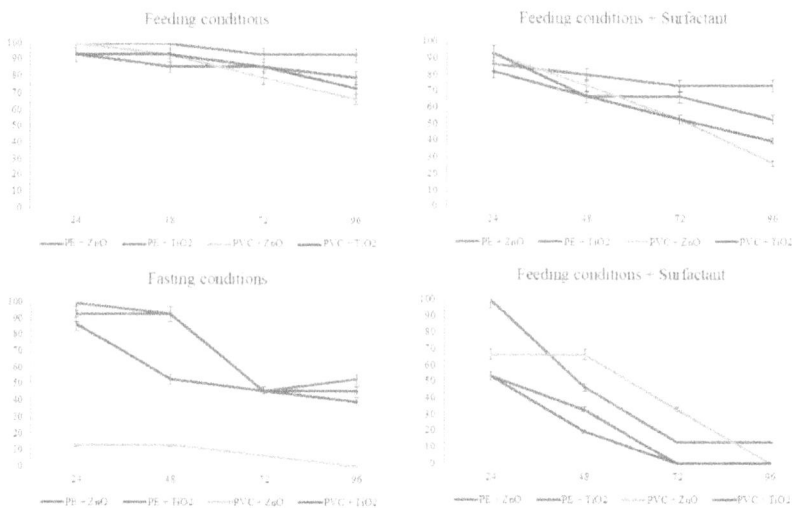

Figure 3.
Contribution of surfactant presence/absence and trophic condition in young organisms. Graphics report the percentages of mobile organisms at the target time normalized for negative controls.

surfactants seem to improve the contact among microplastics and animals and therefore cause effects on exposed specimens [22]. In a previous study [29], exposure to microplastics + surfactant showed the highest toxicity on *D. magna*, also, supporting results obtained on tested complex mixture.

3.2 Feeding effect

The effect of feeding is almost always significant (**Figure 3**) on the toxicity of mixtures of contaminants on *D. magna*. Animals actively ingest nanoparticles during both fast and feeding conditions, but feeding condition reduces the effects of mixtures on the exposed animals. In young organisms, the only case in which the effect of food was not significant was due to the exposure to PE in the presence of n-TiO$_2$.

In the case of adult individuals (**Figure 2**), the effect of feeding was significant in 5 out of 8 cases: in each experiment for PE+n-TiO$_2$, partially following to exposure to Zn (with PE and PVC always at 96 hours) and with PVC+n-TiO$_2$ at 24 hours.

It can be hypothesized that with the intake of food (phytoplankton) there is an increase in the metabolic rate, resulting in a better chance of detoxifying by *D. magna*, making the effects of toxic mixtures less evident. Concerning n-ZnO, another reason may be that, as the toxicity of metal-oxide nanoparticles reasonably due to the release of metal ions in water [33] some water chemical parameters, such as increasing pH and DOM (occurring during feeding with phytoplankton) could have reduced the concentration of free Zn^{2+} released from n-ZnO. Feeding could have affected these key parameters (DOM and pH in particular) following the addition of an external organic source (DOM) and due to changes induced by photosynthesis performed by algal activity on water chemical features as, also, previously hypothesized for single components of complex mixtures tested in this study [28].

3.3 Aging effect

The results achieved provides evidence that the level of toxicity of NPs+MPs mixtures depends on the age of the animals, confirming what yet seen following tests with several kinds of MPs [29]. The effect of aging (comparison between experiments carried out on young people and adults, **Figure 2**) was significant (affecting the possibility of remaining mobile or alive) in half of the experiments carried out. While results at 24, 48 and 72 hours appear controversial, the significance of a toxic effect clearly emerges at 96 hours, both for PE+n-ZnO and for PE+n-TiO$_2$. Significant aging effect also for PE+n-TiO$_2$ at 96 hours and PVC+n-TiO$_2$ at 96 hours have been registered.

4. Conclusions

The results obtained in this study made it possible to identify the mixture PE+n-TiO$_2$ as the most toxic following exposure of *D. magna*, both in the presence of surfactant and without, both under fasting and feeding conditions, in the adults and young specimens tested. The variability and, in many cases, the lack of a univocal trend in the toxicity effects observed between mixtures of different compositions, is reasonably attributable to additive effects to be studied further. On the other side, data that emerge and substantially confirm observations previously performed by research on exposure to single toxicants are: i) the influence of fasting or feeding conditions on toxicity of tested mixture; feeding specimens suffer less from the effects of exposure to contaminants than others; ii) the presence of surfactant (Triton X-100) increases the toxic effects of tested mixtures of contaminants;

iii) adult specimens showed a different resistance to the effects of exposure to mixtures of contaminants than younger specimens (target time 96 hours). Results obtained in this study, furthermore, highlight as interaction between surfactant and other chemicals/materials could induce ecotoxicological responses that cannot be predicted only based on single component tests. This effect is particularly relevant in the real world when animals are exposed under feeding conditions and for longer exposure time than during standardized tests.

Acknowledgements

Authors are grateful to BsRC research centre (Italy) that founded and completely supported this research. Funding: This work was supported by Bioscience Research Center [grant number RG2020009].

Abbreviations

MWWTP	municipal wastewater treatment plants
MP	microplastic
NP	nanoparticle
TiO_2	titanium dioxide
ZnO	zinc oxide
REACH	European Regulation on Registration, Evaluation, Authorisation and Restriction of Chemicals
OECD	organization for economic co-operation and development
PE	polyethylene
PVC	polyvinyl chloride
μFT-IR	micro Fourier transform interferometer
ATR	attenuated total reflection
DOM	dissolved organic carbon

Author details

Cristiana Guerranti[1], Serena Anselmi[2], Francesca Provenza[1,2], Andrea Blašković[3] and Monia Renzi[1*]

1 Department of Life Science, Università degli studi di Trieste, Via Licio Giorgieri, Trieste, Italy

2 Bioscience Research Center, Via Aurelia Vecchia, Orbetello, Italy

3 Voditeljica projekta MPA ENGAGE, Pula, Croatia

*Address all correspondence to: mrenzi@units.it

IntechOpen

References

[1] Renzi M, Perra G, Guerranti C, Franchi E, Focardi S (2009) Abatement efficiency of municipal wastewater treatment plants using different technologies (Orbetello lagoon, Italy). Int J Environ Health 3(1):58-70.

[2] Renzi M, Giovani A, Focardi SE (2012) Water pollution by surfactants: Fluctuations due to tourism exploitation in a lagoon ecosystem. J Environ Prot 3:1004-1009.

[3] Guerranti C, Martellini T, Perra G, Scopetani C, Cincinelli A (2019) Microplastics in cosmetics: Environmental issues and needs for global bans. Environmental Toxicology and Pharmacology 68:75-79.

[4] Gottschalk F, Sonderer T, Scholz RW, Nowack B (2009) Modelled environmental concentrations of engineered nanomaterials (TiO_2, ZnO, Ag, CNT, fullerenes) for different regions. Environ Sci Technol 43:9216-9222.

[5] Pettitt ME, Lead JR (2013) Minimum physicochemical characterisation requirements for nanomaterial regulation. Environ Int 52:41-50.

[6] Kumar P, Morawska L, Birmili W, Paasonen P, Hu M, Kulmala M, Harrison RM, Norford L, Britter R (2014) Ultrafine particles in cities. Environ Int 66:1-10.

[7] Nowack B, Bucheli TD (2007) Occurrence, behavior and effects of nanoparticles in the environment. Environ Pollut 150:5-22.

[8] Hossain F, Perales-Perez OJ, Hwang S, Román F (2014) Antimicrobial nanomaterials as water disinfectant: Applications, limitations and future perspectives. Sci Total Environ 466-467:1047-1059.

[9] Galgani F, Hanke G, Maes T. (2015) Global distribution, composition and abundance of marine litter, in: Bergmann M, Gutov L, Klages M (Editors) Marine Anthropogenic Litter, Springer, Berlin pp. 29-57

[10] Perosa M, Guerranti C, Renzi M, Bevilacqua S (2021) Taking the sparkle off the sparkling time. Mar Pollut Bull 170: 112660.

[11] Tagg AS, Ivar do Sul JA (2019) Is this your glitter? An overlooked but potentially environmentally-valuable microplastic. Mar Pollut Bull 146:50-53.

[12] Kim D, Chae Y, An Y-J (2017) Mixture toxicity of nickel and microplastics with different functional groups on *Daphnia magna*. Environ Sci Technol 51:12852-12858.

[13] Cole M, Lindeque P, Fileman E, Halsband C, Goodhead R, Moger J, Galloway T (2013) Microplastic ingestion by zooplankton. Environ Sci Technol 47:6646-6655.

[14] Fossi MC, Coppola D, Baini M, Giannetti M, Guerranti C, Marsili L, Panti C, de Sabata E, Clò S (2014) Large filter feeding marine organisms as indicators of microplastic in the pelagic environment: The case studies of the Mediterranean basking shark (*Cetorhinus maximus*) and fin whale (*Balaenoptera physalus*). Mar Environ Res 100:17-24.

[15] Renzi M, Blašković A, Bernardi G, Russo GF (2018b) Plastic litter transfer from sediments towards marine trophic webs: A case study on holothurians. Mar Pollut Bull 135:376-385.

[16] Renzi M, Guerranti C, Blašković A (2018) Microplastic contents from maricultured and natural mussels. Mar Pollut Bull 131:248-251.

[17] Wright SL, Thompson RC, Galloway TS (2013) The physical impacts of

microplastics on marine organisms: A review. Environ Pollut 178:483-492.

[18] Talvitie J, Mikola A, Koistinen A, Setälä O (2017) Solutions to microplastic pollution – Removal of microplastics from wastewater effluent with advanced wastewater treatment technologies. Water Res 123:401-407.

[19] Ziajahromi S, Neale PA, Rintoul L, Leusch FDL (2017) Wastewater treatment plants as a pathway for microplastics: Development of a new approach to sample wastewater-based microplastics. Water Res 112:93-99.

[20] Jing Q, Yi Z, Lin D, Zhu L, Yang K (2013) Enhanced sorption of naphthalene and p-nitrophenol by nano-SiO_2 modified with a cationic surfactant. Water Res 47:4006-4012.

[21] Lechuga M, Fernández-Serrano M, Jurado E, Núñez-Olea J, Ríos F (2016) Acute toxicity of anionic and non-ionic surfactants to aquatic organisms. Ecotoxicol Environ Saf 125:1-8.

[22] Frydkjær CK, Iversen N, Roslev P (2017) *Daphnia magna*: Effects of regular and irregular shaped plastic and sorbed phenanthrene. Bull Environ Contam Toxicol 99:655-661.

[23] Bakir A, Rowland SJ, Thompson RC (2014) Enhanced desorption of persistent organic pollutants from microplastics under simulated physiological conditions. Environ Pollut 185:16-23.

[24] Schwarzenbach RP, Escher BI, Fenner K, Hofstetter TB, Johnson CA, von Gunten U, Wehrli B (2006) The challenge of micropollutants in aquatic systems. Science 313(5790):1072-1077.

[25] Renzi M, Guerranti C (2015) Ecotoxicity of nanoparticles in aquatic environments: A review based on multivariate statistics of metadata. J environ Anal Chem 2(4):149.

[26] Baird DJ, Barber I, Bradley M, Calow P, Soares AMVM (1989) The Daphnia bioassay: A critique. Hydrobiologia 188:403.

[27] OECD (Organization for Economic Cooperation and Development testing guidelines) guidelines for Daphnia species, acute immobilization tests: OECD guideline n. 202, 2004.

[28] Renzi M, Blašković A (2019) Ecotoxicity of nano-metal oxides: A case study on *Daphnia magna*. Ecotoxicology 28:878-889.

[29] Renzi M, Grazioli E, Blašković A (2019) Effect of different microplastic types and surfactant-microplastic mixtures under fasting and feeding conditions: A case study on *Daphnia magna*. Bulletin of Environmental Contamination and Toxicology 103:367-373.

[30] Baumann J, Sakka Y, Bertrand C, Köser J, Filser J (2014) Adaptation of the Daphnia sp. acute toxicity test: miniaturization and prolongation for the testing of nanomaterials. Environ Sci Pollut Res 21(3):2201-2213.

[31] Khoshnood R, Jaafarzadeh N, Jamili S, Farshchi P, Taghavi L (2016) Nanoparticles ecotoxicity on Daphnia magna. Transylv rev Syst Ecol res "Wetl Divers" 18(2):26-32.

[32] Wypych G (2015) PVC properties. In: PVC formulary (2nd Ed.), pp. 5-44. DOI:10.1016/B978-1-895198-84-3. 50004-1

[33] Blinova I, Ivask A, Heinlaan M, Mortimer M, Kahru A (2010) Ecotoxicity of nanoparticles of CuO and ZnO in natural water. Environ Pollut 158:41-47

Surfactants and Their Applications for Remediation of Hydrophobic Organic Contaminants in Soils

Roger Saint-Fort

Abstract

Soil contaminated with ubiquitous hydrophobic organic contaminants (HOCs) is a worldwide recurring concern arising from their indiscriminate disposal, improper management, and accidental spills. A wide range of traditional remedial strategies have been the common practice. However, these treatment methods have become cost prohibitive, not environmental friendly, and less accepted by society. Surfactant-enhanced remediation technology represents a cost-effective and green technology alternative to remediate such contaminated sites. Surfactant remediation technologies are conducted in-situ or ex-situ as two broad categories, or in combination. Among these technologies are soil flushing, washing, phytoremediation, and bioremediation. More applied research continues to quantify the efficiency of surfactant-enhanced mass transfer phase using a single surfactant solution while their binary blends to remove mixed HOCs in soils are also a focus of interest for research. There is a great potential to develop novel synthetic and biosurfactants that will exhibit higher biodegradability, less toxicity, higher removal efficiency, more economical and more recyclable. This work thus provides a review of the applications and importance of surfactant-enhanced remediation of soil contaminated with HOCs. Relevant environmental factors, soil properties, surfactant chemistry, mechanisms, mass transfer phase, and field designs are summarized and discussed with purposes of providing greater context and understanding of surfactant-enhanced remediation systems.

Keywords: Remediation, surfactants, soil, hydrophobic, contaminants

1. Introduction

A major environmental concern around the world is soil contamination by ubiquitous hydrophobic organic contaminants (HOCs) due to their improper management. Such contaminants pose serious environmental and health risks to the public, and can be difficult to remediate due their intrinsic complexity and their weathering. Soils contaminated with HOCs not only can be deleterious to the ecosystem, it can lead to increasing economic loss and ecological insecurity. HOCs which are largely organic in nature, are characterized by relatively low solubility, a specific density that can be greater or less than 1, nonpolar compounds and have been shown to be toxic, mutagenic and/or carcinogenic even at trace concentrations in the soils. Example of HOCs include aromatic compounds in petroleum and fuel residue, chlorinated compounds in commercial solvents, pharmaceutical chemical wastes like trichlorophenol, polycyclic aromatic hydrocarbons (PAHs), polychlorinated biphenyls (PCBs), dichlorodiphenyltrichloroethane (DDT) etc. Furthermore, many HOCs in the soil can be volatile and their behavior may engender vapor intrusions in various structures [1]. As previously reported [2] their availability for biochemical transformations is significantly affected by their large octanol–water partition coefficients (log $K_{ow} > 2$). Even at very low concentrations, HOCs have shown to enter the food chain through various pathways and as bioaccumulating compounds, may ultimately threaten human life and other ecological receptors. Removal of HOC in soils can represent a significant challenge because such efforts can be site specific, costly, and often with limited success for its associated plumes [3]. Particular attentions to the ubiquitous deployment of surfactant-based remedial technologies indicate their ability to provide the means of great practical importance for implementing environmentally friendly remedial solutions, at low cost, and in a scientifically and engineering sound manner. Traditional framework in using surfactant remediation technologies are in-situ or ex-situ as two broad categories, or in combination. Among these technologies are soil flushing, washing, and bioremediation.

The in-situ remedial method involves remediation of the contaminated soil matrix without excavating the contaminated soil. This approach is generally considered less disruptive to the land ecosystem, may require multi-stage of operation, highly affected by soil physical properties and characteristics, and the time required to achieve the remediation effect may be substantial. The long treatment time associated with in-situ remediation may make the site unusable during the remediation period. Several in-situ remediation techniques have been developed which include surfactant aided flushing techniques. In conducting in-situ soil flushing (i.e., soil washing) remediation, a low concentration of surfactant solution is passed through an in-place contaminated soil using a vertical injection or infiltration process. The surfactant solution entrains the dissolved contaminants to an area where it can be collected and removed for treatment or disposal. However, the groundwater beneath the contaminated soil may serve as the discharge point for the extraction fluids. In such instance, the groundwater needs to be treated to adhere to environmental standards and maintain strict environmental quality at the site to protect public health and ecological receptors. Following HOCs in-situ surfactant-enhanced mass transfer phase into the soil solution, phytoremediation has been applied to extract, sequester, and detoxify the contaminants [4]. Since phytoremediation capacity is species specific, using a combination of plants as remedial agents will increase the efficacy of the remedial process. A notable advantage of phytoremediation, it is scientifically referred to as green technology and low cost. However, the time required to achieve the remedial target is typically longer compared to the other in-situ remedial approaches. Surfactants are also used in performing in-situ bioremediation of HOCs. The intended goal is to increase the bioavailability of the organic contaminants through mass transfer dissolution into the soil solution matrix and direct aqueous solubility. In this review, bioremediation is defined as a process, which relies on biological mechanisms to degrade, detoxify, mineralize or transform concentration of organic contaminants to an innocuous state. Often, nitrogen and phosphorous are limited as key soil nutrients and need to be added to biostimulate the soil natural microbial biodegraders. Both phytoremediation and bioremediation in-situ techniques will be affected by climatic conditions at the site.

The ex-situ approach can be conducted on-site or off-site. It involves excavating, storing and pre-treating the contaminated soil. Then, followed by treatment and redisposition of the clean soil. Treatment aided surfactant may take place in a variety of ways. Most common approaches involve biopile, windrow, and bioreactor. Such ex-situ practices are more preferable compared to the popular dig and dump method in which the contaminated soil is excavated and dumped in an industrial landfill. Under this widely practiced conventional approach, the contaminants are not mineralized nor destroyed and represent long term threat to human health and ecological systems [5]. Environmentally friendly and cost saving features are among the major advantages of surfactant-enhanced bioremediation offer compared to landfilling, chemical and physical methods of remediation. However, the higher costs associated with transportation and associated liabilities of moving hazardous soil, and destruction of the soil ecosystem associated with excavation summarize the main disadvantages of soil ex-situ remediation over in-situ. In some instances, ex-situ treatment is preferred as a treatment as it offers more redeployment options of the land, treatment endpoint occurs faster and often the feasibility of being used with other treatment methods.

In their common form, surfactants are a group of amphiphilic chemicals constituted by both a hydrophobic moiety (chain) and a hydrophilic moiety (head) in the molecular structure of varying length in various surfactants. In fact, the unusual properties of aqueous surfactant solutions are best ascribed to the presence of the polar or ionic head group that interacts with an aqueous environment which leads to the solvation of the surfactant via ion-dipole or dipole–dipole interactions. Surfactants (short for "surface active agents" represent a unique class of compounds with distinct chemical and physical properties. Surfactants unique molecular structure give them the ability to dramatically alter interfacial and surface properties as well as to self-associate and solubilize themselves in micelles [6].

Surfactants manufactured by petrochemical plants are known as synthetic or chemical surfactants. Those produced from biological organisms are known as natural or biosurfactants. These lead to a vast array of their practical applications in terms of health, care products, food, petroleum processing etc. Irrespective of their source, the hydrophilic head group in the surfactant molecule is considered to be the main factor responsible for their special chemistry [7]. Historically, the costs of synthetic surfactants production remain comparatively less than biosurfactants. Several health and environmental concerns arise from using petroleum-based surfactants. In this regard, they are marginally biodegradable, can pollute soil and water, may bioaccumulate in the environment, and disruption of the endocrine system. On the other hand, biosurfactants being derived from biotechnology processes, are more environmentally friendly substance and often referred to as green technology. However, like petroleum-based surfactants, natural surfactants are associated with skin irritation and allergies. Considering the vast array of surfactants molecular structure and properties, one can anticipate an increase use in a myriad of environmental application for decontamination of soil matrices. This entails that surfactants with different properties and molecular structures can be strategically selected for different soil decontamination purposes. Importantly, particular consideration should be given to determine combining various surface agents for achieving greater remedial efficiency. This work provides an examination of surfactant-enhanced remediation of soil contaminated with hydrophobic organic contaminants as well as practical and general considerations involved in their implementation.

2. Classification of surfactants

Surface active compounds are the most commonly used chemicals in everyday life. The number of different molecules of surfactants that have been manufactured must be in in the thousands and many have found practical use in society.

Unfortunately, it is somewhat surprising that surfactants, until only very recently, been explored for environmental remediation applications. Architecturally, a surfactant molecule contains a chain, the hydrophobic moiety, that can be linear or branched while the head is the polar or ionic moiety [1] (**Figure 1**). The hydrophobic is typically a hydrocarbon chain of an average of length of 12 to 18 carbon atoms and may involve an aromatic ring. For the purpose of this review, surfactants are divided into four main categories depending on the nature of the polar moiety as depicted in **Table 1** [1]. An in-depth discussions of surfactants chemistry and structure are presented elsewhere [6–8]. Furthermore, there are a number of review of publications available for surfactants use in specific industries [9]. A summary of basic information of various surfactants that have been used for the remediation of soil contaminated with HOCs is depicted in **Table 2**.

Biosurfactants are a group of surface active agent biomolecules produced by microorganisms. It has been suggested that surface active biomolecules can be best divided into low-molecular mass molecules or higher-molecular mass polymers. An adaptation of their classification is depicted in **Table 3**.

In recent years, scientists have been working diligently at evaluating the effectiveness of various types of surfactants to degrade organic contaminants in soils. In

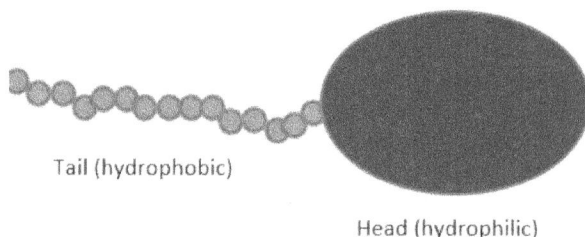

Figure 1.
Structural parts of conventional surfactant molecule.

Head charge/chemical structure example	Group class
sodium octyl sulfate	Anionic
Cetrimonium bromide	Cationic
3-[(3- cholamidopropyl) dimethylammonio]-1-propanesulfonate.	Zwitterionic or Ampholytic
Span 80	Nonionic

Table 1.
Category of surfactants classification [1].

Surfactant	Name/components	Type	Molecular formula	MM (g/mol)
TX100	P-tertiary-octylphenoxy polyethyl alcohol	Noninionic surfactant	$C_{14}H_{22}O(C_2H_4O)_n(_{n = 9-10})$	625
CAPB	Cocoanut amide propyl betaine	Zwitterionic surfactant	$C_{19}H_{38}N_2O_3$	342.52
SDS	Sodium dodecyl sulfate	Anionic surfactant	$NaC_{12}H_{25}SO_4$	288.372
AOS	Alpha olefin sulfonate	Anionic surfactant	$C_nH_{2n}\text{-}1SO_3Na$ (n = 14–16)	324
SLES	Sodium laureth ether sulfate	Anionic surfactant	$CH_3(CH_2)_{10}CH_2(OCH_2CH_2)_nOSO_3Na$	288.38
Tween 80	Polyoxyethylene sorbitan monooleate	Nonionic surfactant	$C_{64}H_{124}O_{26}$	1310
Surfactin	Cyclic lipopeptide	Zwitterionic biosurfactant	$C_{53}H_{93}N_7O_{13}$	1036.3
Brij 35	Poly(oxyethylene)$_{23}$ dodecyl ether	Nonionic surfactant	$C_{12}H_{25}(OC_2H_4)_{23}OH$	1198
Saponin	Pentacyclic triterperne saponin	Nonionic biosurfactant	$C_{58}H_{94}O_{27}$	1223.3
Sophorolipid	Sophorolipid	Nonionic biosurfactant	$C_{34}H_{58}O_{15}$	706.8
Tergitol NP-10	Polyethylene, mono (p-nonylphenyl) ether	Nonionic surfactant	$C_{15}\text{-}H_{24}\text{-}O(C_2\text{-}H_4\text{-}O)n$	642 (average)
Calfax 16 L-35; Dowfax 8390	Sodium heaxadecyldiphenyl ether disulfonate	Anionic gemini surfactant	$C_{28}H_{40}Na_2O_7S_2$	598.72
CAHS	Cocamydopropyl hydroxysultaine	Zwitterionic surfactant	$C_{20}H_{42}N_2O_5S$	422.62
APG	Alkyl polyglucosides	Nonionic biosurfactant	$C_nH_{2n}O_6$	320–370

Table 2.
Basic information of surfactants used in soil remediation of HOCs.

this section, the classification and discussion of surfactants will be more specifically focused on surfactants that have practical relevance in the remediation of soil contaminated with HOCs.

2.1 Ionic surfactants

The family of ionic surfactants is comprised of cationic, anionic and zwitterionic surfactants. They have been applied successfully for the mass transfer solubilization and removal of a variety of HOCs such as PCBs, dense nonaqueous phase liquid (DNAPLs), light NAPLs, BTEX in different soil types. Many literature documents their success in laboratory scale testing and from site-specific soils at pilot or full scale [8–12]. Interest in developing more effective (higher performance/cost ratio) and less toxic surfactants formulation has led to the emergence of Gemini surfactants. It has been reported that the surface active of Gemini surfactants could be of a several order of magnitude greater than conventional surfactants [13]. They are

Biosurfactants class		
Microorganisms origin		Photogenic origin
Low molecular mass	High molecular mass	
Glycolypids: Conjugates of fatty acids and carbohydrates. Most common biosurfactants: trehalopids, Sophorolipids, rhamnolipids. *Burkholderia plantarii.* Producing microorganisms: *Mycobacterrium, Arthrobacter spp, Pseudomonas aeruginosa*	**Polymeric biosurfactants:** Typically consists of three to four Repeating sugars with fatty acids attached to them. Most common biosurfactants: emulsan, liposan, alasan Producing microorganisms: *acinethobacter calcoaceticus, candida lipolytica*	Saponins, lecithins, soyprotein, lactonic, soybean oil, glycolipid, Sunflower seed
Lipopeptides and lipoproteins: Consist of a lipid attached to a polypeptide chain. Most common biosurfactants: surfactin and lichensyn Producing microorganisms: *Bacillus sp.*	**Particulate biosurfactants:** Can be extracellular vesicles and whole microbial cell. Most common biosurfactants: vesicles, whole microbial cells. Producing microorganisms: *acinetobacter calcoaceticus, pseudomonas marginalis, cyanobacteria*	
Phospholipids, fatty acids and neutral lipids: Length of hydrocarbon chain in their structures determines the hydrophilic and hydrophobic balance. Most common biosurfactants: corynomycolic acid, phosphatidylethanolamine Producing microorganisms: *Rhodococcus erythropolis, corynebacterium lepus*		

Table 3.
Biosurfactants classification (adapted with permission from [1]).

used to describe dimeric surfactants which are surfactants that have two hydrophilic (chiefly ionic) polar groups and two hydrophobic chains on each surfactant molecule (**Figure 2**). These twin parts of the surfactant are linked through a spacer of varying link [14]. Gemini surfactants offer a number of superior properties when compared to conventional ionic surfactants. These advantages can be best

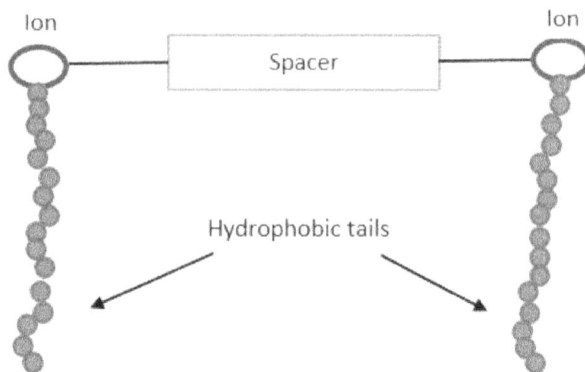

Figure 2.
Illustration of a gemini surfactant.

summarized as lower concentration requirement for solubilizing HOCs, higher aggregation at significantly lower concentration, superior wetting agent, surmount hard-water tolerance effect on mass transfer into soil solution and increased surface activity (C_{20}).

2.2 Nonionic surfactants

Nonionic surfactants are a group of surfactants that hardly dissolve in water, are neutral, and do not have any charge on their hydrophilic end. Their polar portions are typically made up of oxygen-containing groups. Nonionic surfactants solubilize in aqueous phase through hydrogen bonds formation of hydrophilic moieties with water. As the temperature is raised, it reaches the point at which large aggregates of the nonionic surfactant separate out into a distinct phase. There are several properties of nonionic surfactants that make them more suitable candidates to use in soil remediation of HOCs compared to ionic surfactants. Nonionic surfactants tend to have low toxcity, more biodegradable, more cost-effectiveness, low susceptibility to aggregate clay minerals, and low CMC. In the context of this review, toxicity is the measurable adverse effect that a surfactant will have on the soil microorganisms, while biodegradability refers to the ability of the soil microorganisms to destroy the surfactant. The literature abounds with scientific reports that document the wide application of nonionic surfactants for site-specific contaminated by a variety of HOCs [15–18].

2.3 Biosurfactants

Recently, there has been significant research interest in developing and investigating cost-effectiveness production of biosurfactants with unique properties and potential wide applications. One germane challenge that environmental scientists faces in the application of synthetic surfactants-enhanced soil remediation is their toxicity and biodegradability in the environment. It is noted that the environmental applications of biosurfactants has been gaining rapid interest and acceptance in the field of surfactant-enhanced soil remediation. This is due to their attractive physicochemical properties, low toxicity, high biodegradability and relative ease of preparation make these surface active biomolecules suitable candidates for soil remediation.

3. Classification of soil matrices

3.1 Laboratory method

Investigation of contaminated soils requires determining their physical properties for their classification. To this effect, soil classification can be approached from the perspective of the soil texture and organic matter content. The co-influence of both characteristics will have significant impact on the behavior of contaminants and surfactants when performing surfactant-enhanced soil remediation. Such impact is demonstrated through sorption and desorption, bioavailability, mechanism of interactions, contaminants leaching and fate in the soil. Most soils consist of a combination of sand, silt and clay and their range in size is reported in **Table 4**. Depending how much clay, sand, and silt that are present, the soil is given a name. The textural class of a soil is determined by the percentage of sand, silt, and clay. Soil texture determination begins by segregating the fine earth from the rock fragments. Fine earth refers to soil fraction that passes through a #10 sieve. It includes

Particles name	Particle diameter (mm)
Very coarse sand	2.0 to 1.0
Coarse sand	1.0 to 0.50
Medium sand	0.50 to 0.25
Fine sand	0.25 to 0.10
Very fine sand	0.10 to 0.05
Silt	0.05 to 0.002
Clay	< 0.002

Table 4.
Size range of soil particles.

all particles smaller than 2 mm in diameter. Sand, silt, and clay particles are components of fine earth. These three are called the separates of the fine earth. The soil textural triangle (**Figure 3**) is a representation of the mineral content of a soil in various combinations of clay, silt, and sand. The most common method for determining soil texture is the hydrometer method. According to this method, the soil separates are dispersed with solution of sodium metaphosphate (Calgon), blended and the density of the suspension measured at various time intervals. After dispersion, the amount of each particle group (sand, silt, clay) are determined by using a hydrometer. Once these percentages have been determined by the hydrometer method, the triangle can then be used to determine the soil textural class name.

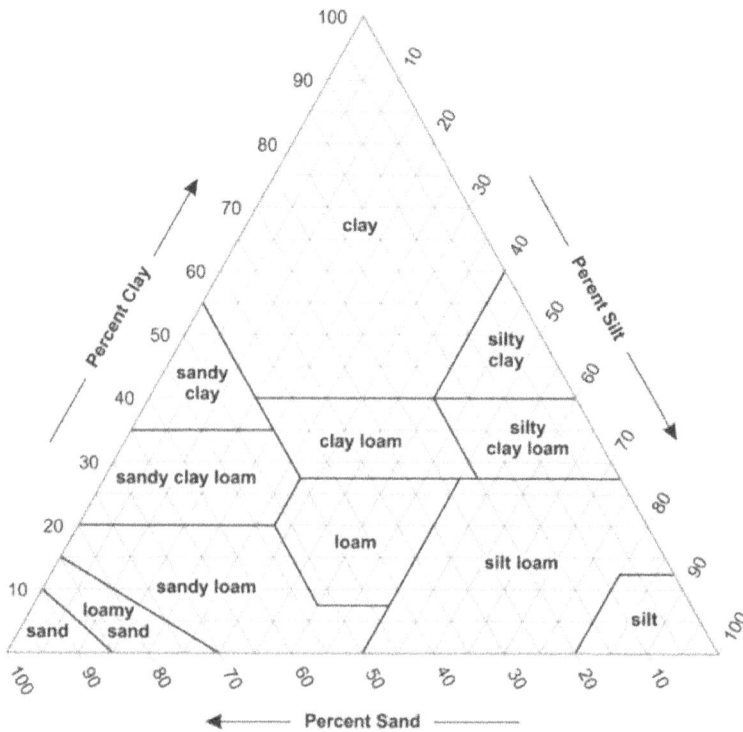

Figure 3.
USDA triangle representation of textural soil classes.

3.2 Field estimate assessment

A rough estimate of a soil textural class can be obtained by the method of feel. This method is used by environmental scientists and engineers in preliminary site reconnaissance, detailed site and contaminant characterization, sampling for transport and fate modeling, risk assessment, and in remediation selection and design. Development and execution of textural field program is relatively simple and inexpensive. In conjunction, overall project costs may be reduced as field method provides a more efficient alternative to other more complex and expensive methods. However, when dealing with a contaminated site, safety requires that one should wear gloves and avoid direct contact with contaminated soil material being assessed.

The method of feel is based on visual and tactile observations. This technique involves working a wet soil sample between the thumb and fingers to estimate the amount of sand, silt, and clay. Rarely, if ever, does a particular soil consists wholly of one soil separate or size fraction. General properties of the three major soil separates are reported in **Table 5**. The method by feel requires some practice to acquire a high level of proficiency.

Soil separate	Diameter of particles	General characteristics
Sand	2–0.05 mm	Individual particles feel gritty when the soil is rubbed between fingers. Not plastic or sticky when moist. Moist sample collapses after squeezing.
Silt	0.05–0.002 mm	Feels smooth like flour or corn starch and powdery when rubbed between the fingers. Not plastic or sticky when moist.
Clay	Less than 0.002 mm	Feels smooth, sticky, and plastic when moist. Forms very hard clods when dry. Particles may remain suspended in water for a very long period of time.

Table 5.
Basic characteristics of soil separates.

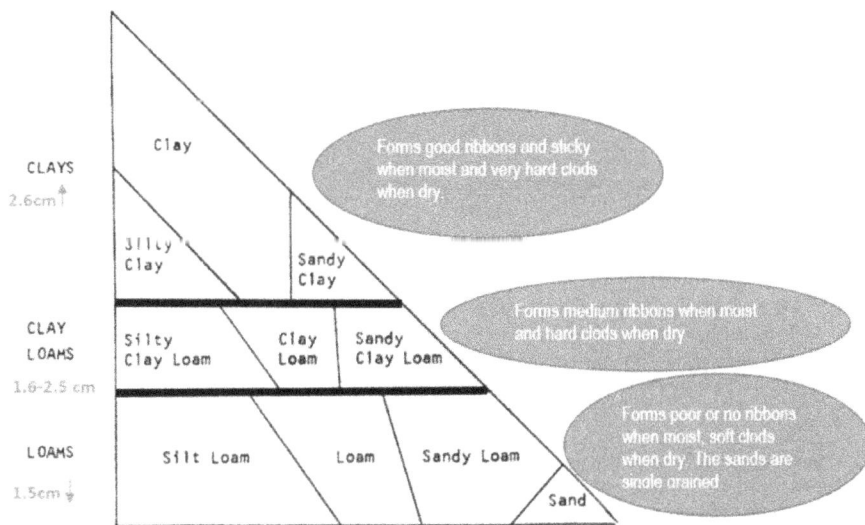

Figure 4.
Modified textural triangle for determining soil texture by the feel method.

3.2.1 Modified textural triangle

It can be used on contaminated soils when the conditions and context favor its use. Clay content is estimated by the length of the soil ribbon formed and is referred in **Figure 4**. Both, **Figure 4** and **Table 5** can be used to estimate the textural class name for a contaminated soil.

3.2.2 The unified soil classification system

The Unified Soil Classification System (USCS) established in 1942 is also commonly used in contaminated site environmental investigation. Each type of soil is given a two-letter designation based on its texture, atterberg limit and organic matter content (**Table 6**). Since every soil contains a blend of soil separates, the

USDA modified textural triangle	USCS letter symbols
Loam:	**ML:**
• Has a good blend of silt and clay, moderate looseness • Forms short ribbons • Sand is noticeably felt but does not dominate	As per Loam clues
Sandy loam:	**SM:**
• Forms very poor ribbons, good looseness • Enough silt and clay to give the sample body • Sand presence dominates • Moist soil does not crumble entirely after squeezing	As per Sandy Loam clues
Silt loam:	**ML:**
• Forms short, broken ribbons, average to moderate looseness • Not sticky when moist • Fells smooth, powdery	As per Silt Loam clues
Clay loam:	**CL:**
• Forms medium, broken ribbons, poor looseness • Noticeably plastic and sticky • A great amount of girt • Relatively hard to work between thumb and forefinger	As per Clay Loam clues
Sandy clay loam:	**SC:**
• Forms short medium, broken ribbons, moderate looseness • Feels gritty and sticky	As per Sandy Clay Loam clues
Silty clay loam:	**MH:**
• Forms high to medium, broken ribbons, moderate looseness • Feels smooth and sticky • Does not feel very gritty	As per Silty Clay Loam clues
Sandy clay:	**MH or CH:**
• Forms relatively long, broken ribbons, poor to average looseness • Feels unequivocally sandy and vary gritty • Sample very hard to work between thumb and forefinger • A lot of water is needed to wet a dry sample before it can be worked	As per Sandy Clay clues
Silty clay:	**CH:**
• Forms relatively long, broken ribbons, poor looseness • Feels very velvet smooth, very sticky, hard to break • Dry sample takes a lot of water to wet	As per Silty Clay clues

Table 6.
Tactile and observational clues related to textural classes for USCS and USDA.

possibility of soil that contains only sand or clay is not realistic. For additional information on soil classification by feel, the reader is referred elsewhere [19, 20].

4. Behavior of surfactants at soil/liquid interface

Surfactants at very low concentration can solubilize HOCs by reducing surface and interfacial tensions of the soil water solution. Surfactants will typically consist of a strongly hydrophobic group (water hating) referred to as the tail of the molecule and a strongly hydrophilic group (water loving), which is the head. Owing to the hydrophilic portion, surfactants can exhibit high solubility in water, while the hydrophobic portion causes part of the molecule to reside in an insoluble phase. Hydrogen bonding property and Van deer Waals forces between water molecules are the main reasons for preventing HOCs to form aqueous solutions in a soil system. Therefore, their mass distribution is primarily confined to the solid phase of the contaminated soil. However, at a specific, higher concentration of surfactant, commonly known as the critical micelle concentration (CMC), molecular aggregates are formed. The CMC is a specific property of a surfactant. In technical term, the CMC value represents the concentration of maximum solubility of a surfactant at 25°C in a particular aqueous soil solution. It should be noted that the effectiveness of CMC at a contaminated site may be affected by temporal and seasonal variations exhibited by the soil solution properties. It is through micellar solubilization, the process by which aggregations of surfactant monomers form micelles that HOCs canbecome solubilized. The solubilization process dictates the suitable approach in relation to remedial options and site-specific characteristics. The presence of surfactants in the soil solution will be accompanied by an interplay between the soil solution and concentration of surfactant. An adaptation of the interplay is depicted in **Figure 5**. Therefore, surface activity of surfactants should be viewed as a dynamic phenomenon. The solubilization of HOCs in the soil solution is accompanied by an increase in the Gibbs energy transfer which results in a decrease in entropy. This thermodynamic process is believed to to be the result of the breakdown of hydrogen bonding in the water molecule. Generally, the lower the CMC of a surfactant molecule in a soil system, the more stable will be the micelles and correspondingly the mass transfer process. The most commonly held view of key

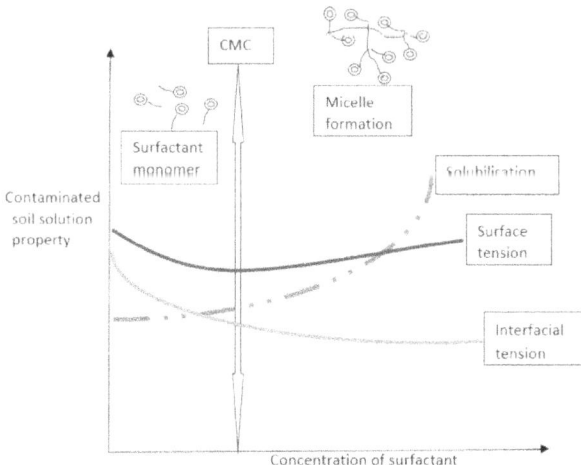

Figure 5.
Illustration of various interplays at the soil water-interface and HOCs. (Reproduced with permission from [1].)

factors affecting micellar solubilization of HOCs in soil by nonionic, ionic, and biosurfactants are the following: soil moisture, sorption, soil moisture, salinity, surfactant hydrophobic properties, texture, organic carbon, pH, and interfacial energy [1].

The effectiveness of a particular surfactant in solubilizing a specific HOC can be determined through the molar solubilization ratio (MSR) and micelle-water partition coefficient (K_{mc}). The MSR is the number of solute molecules solubilized per surfactant molecule. Namely the MSR can be calculated according to Eq. (1):

$$MSR = (S–S_{CMC})/(C_s–CMC) \tag{1}$$

where

MSR = moles of organic contaminant solubilized per mole of surfactant added to the aqueous phase

S = apparent solubility of organic contaminant at a given surfactant concentration.

S_{CMC} = CMC point of surfactant.

C_s = apparent solubility of organic contaminant at CMC (i.e., C_s > CMC).

CMC = critical micelle concentration

Studies on mixed surfactant systems competitive effects on hydrophobic contaminants solubilization has been investigated and reported elsewhere [21–23]. In mixed surfactants, the MSR for the HOC can be estimated using the MSR obtained in single-surfactant solutions assuming the ideal mixing rule [24] and can be represented by Eq. (2):

$$MSR_m = Y_1MSR_1 + Y_2MSR_2 \tag{2}$$

where

MSR_m = moles of surfactant solubilized in mixed surfactants

Y_1 and Y_2 = molar fractions of the two surfactants

MSR_1 and MSR_2 = molar solubilization ratios for the HOC

A plot of the aqueous HOC concentration solubility versus surfactant concentration, MSR and K_{mc} can be determined from the slope of the linearly fitted regression equation, respectively.

The K_{mc} can be obtained from Eq. (3):

$$K_{mc} = \frac{MSR}{(1+MSR)V_w\,S_{CMC}} \tag{3}$$

where the variables are as previously defined.

It is suggested that the greater the values of MSR and K_{mc} the larger the solubilization capacity of the surfactant in the soil micellar solution.

The micelle-aqueous phase partition coefficient (K_m) is often used as another approach to quantify the solubilization capacity of a single surfactant [14]. Eq. (4) can be used to obtain K_m:

$$K_m = X_m/X_a \tag{4}$$

where

X_m = the mole fraction of hydrophobic compounds encapsulated in the micellar phase given by {MSR / (1+ MSR)}.

X_a = the mole fraction of hydrophobic compounds in the aqueous phase

The soil-water partition coefficient K_d is a parameter commonly used to determine the relative affinity of a contaminant for the solid phase, C_s, and aqueous phase, C_w. The greater the K_d value means that a contaminant tends to accumulate onto the soil matrix. K_d can be obtained from Eq. (5):

$$K_d = \frac{C_s}{C_w} \tag{5}$$

The apparent soil-water partition,, K'_d, can be determined from adsorption equilibrium and we get Eq. (6):

$$K'_d = \frac{K_d + C_{sorbed}\, K_{psf}}{1 + C_{micelle}\, k_{mc}} \tag{6}$$

where
C_{sorbed} = the amount of surfactant sorbed onto the soil
K_{psf} = the partition coefficient of the HOCs in the sorbed surfactant
$C_{micelle}$ = concentration of micelle in soil solution
K_{mc} = micelle-water partition coefficient
For in-situ soil washing and surfactant-enhanced bioremediation, the solubilization potential of the HOC should be optimized. Basic information on the soil properties regarding range and distribution pattern of pH, texture, organic carbon, and salinity should be determined. Strategic adjustments in the delivery and concentration of the surfactant solution can be made.

5. Environmental risks and toxicity of surfactants

Surfactants are economically important and vital to our economy. They are a diverse group of chemicals, widely used by society and continue to be part of our daily life. However, as new surfactants are synthesized annually and surfactants production overall continue to rise, concerns about their impact on the environment and human health have been raised and studied [25–27]. Achieving high contaminant removal and mass transfer without causing any negative effects on the soil system are the primary considerations in the application of surfactants. Toxicity and biodegradability of surfactants are typically tested under different environmental conditions based on the intended application. Typically, most surfactants are not considered acutely toxic to organisms at concentrations typically encountered in the environment. Toxicity is measured in terms of effective concentration (EC_{50}) or lethal concentration (LC_{50}). EC_{50} represents the surfactant concentration (mg L^{-1}) that results in a 50% reduction in a microbial population or a biological community. LC_{50} refers to the concentration of a surfactant that causes the death of the microbial soil community or living organisms after 96 hours of exposure. Surfactants, including their metabolites, that have a toxic effect on a soil microbial community is referred to as xenobiotic surfactants. The harmful effects of xenobiotic surfactants occur through the rupture and penetration of the cellular membrane by interacting with lipids and proteins [28]. Nonetheless, the relationship between surfactants chemical structure, physicochemical parameters, biological activity and environmental impact is still ambiguous. Even less studied and understood are the comingling effects of multiple surfactants on the soil ecosystem. It can be hypothesized while a single surfactant may have minimal adverse impact on the environment. In the presence of other surfactants, it may have antagonistic effects in the soil and other terrestrial ecosystems.

In general, the two main challenges related to surfactant-enhanced soil remediation are their toxicity and biodegradability. Surfactants are considered to be biodegradable if its molecular structure can be mineralized by the soil natural microbes through metabolic activities. On the other hand, toxicity reflects the adverse impact created by surfactants on the soil biota. Generally, the order of surfactants toxicity are biosurfactants < nonionic < anionic < cationic. Toxicity effects of surfactants may occur when a surfactant coats, sorbs onto soil particles and accumulate to toxic level. This leads to the formation of a hydrophobic layer around the soil aggregates which modifies the soil hydrophobicity. The effects are destruction of soil ability to absorb water, reduction of water infiltration into the soil. If surfactants accumulate in soils to toxic level around the plants rhizosphere, the phytotoxicity effects of the surfactant will lead to growth reduction and crops yield or death of vegetation. Most synthetic surfactants used in soil remediation are not readily biodegraded by the soil microbes and can result in toxic adverse effects on the soil ecosystem [29]. Ionic surfactants such as SDS and CATB are highly biodegradable, but exhibits high toxicity. In contrast, the nonionic surfactant Tween [30] and the biosurfactant Rhamnolipid [31] are highly biodegradable and has low toxic. Surfactants used in soil remediation and their degradation products may leach into the aquifer or enter other components of the terrestrial system. The endocrine system is a network of glands and organs that produce, store, and secrete hormones. If exposed to these substances, they would have the potential to disrupt the normal functioning of endocrine system in wildlife and human beings [32].

6. Mechanisms of surfactant-enhanced solubilization

The typical soil system will consist of five distinct phases represented by solid, solution, organic, and gaseous. When HOCs are released in a soil system, the natural dynamic processes of immobilization and demobilization, and mobilization processes occur without solubilization enhancement (**Figure 6A**). Immobilization or sorption is the dominant process and implies the removal of HOCs from the soil solution and the soil gaseous phase leading to retention on the soil solid phase. When the system becomes surfactant-enhanced, demobilization or desorption dominates and as a result, the HOC is released from the solid phase into the solution phase (**Figure 6B**). Mobilization or migration which refers to transport of HOCs in the soil porous media is also significantly enhanced by increasing solubilization HOC. These processes interact to influence surfactant-enhanced soil remediation. Studies have shown that mobilization or emulsification and solubilization are the two main mechanisms by which surfactants enhance the mass transfer solubilization of hydrophobic organic contaminants sorbed onto the soil organic matter and sediments in the soil aqueous phase. Mobilization takes place at concentrations of surfactant below CMC, whereas emulsification allows for dispersion of one phase into the other. Surfactant monomers accumulate at the soil-HOC and soil-water interfaces. This has for effect to change the wettability of the soil system by maximizing the contact angle between the HOC and the soil. A repulsion effect between the hydrophobic groups of the surfactant moiety and the rest of the surfactant molecule is caused by surfactant molecules retained on the surface of the HOC, thereby further enhancing the desorption of the contaminant from the soil particles [33]. The solubilization process occurs at concentrations above the surfactant CMC. At the same time, more sorbed HOCs are partitioned in the soil solution phase leading to more contaminant being solubilized and bioavailable. It is inevitable that a certain amount of surfactant will be sorbed onto the soil system and will be ineffective. Sorbed surfactant does not contribute to the mobilization and solubilization dynamism. Mobilization effect results in enhancing soil flushing remediation through transport and leaching of the HOC in the soil porous media and increased

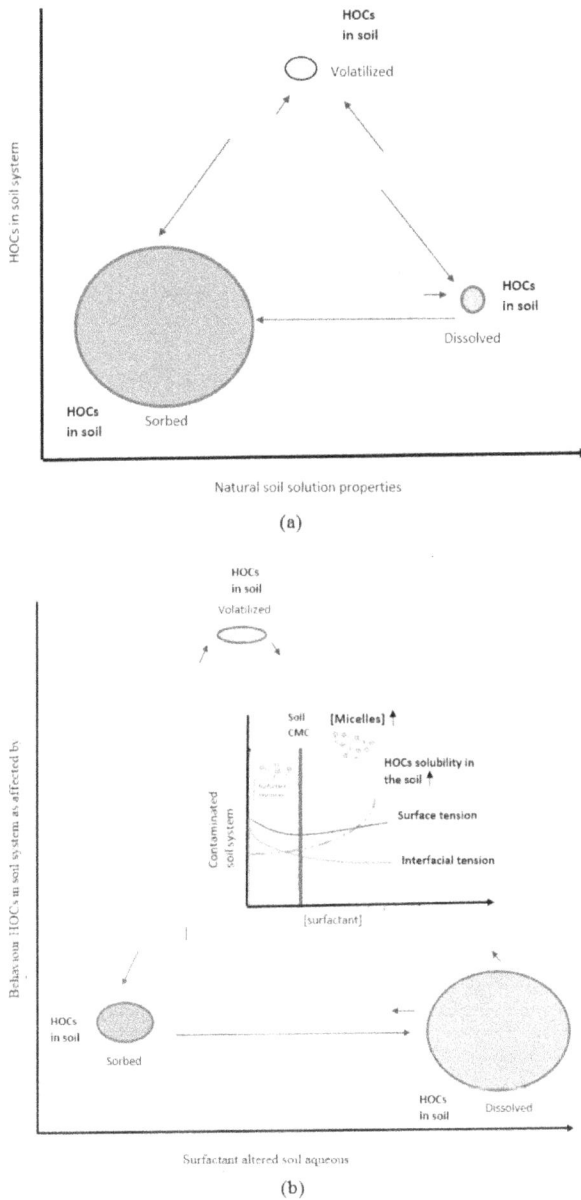

Figure 6.
Illustration of the enhancement effects of surfactant on the solubility of HOCs in a soil system: (A) no surfactant-enhancement, (B) surfactant-enhancement effects. (Arrows indicate intensity of equilibrium between HOC phases).

bioavailability for biodegradation. Basically, solubilization effect of surfactant increases the apparent solubility of HOCs in a contaminated soil.

7. Field strategy and design of surfactant-enhanced remediation

It is important to characterize and delineate the HOC in the soil in order to successfully implement a surfactant-enhanced remedial program. In this review, a simplified overview of the main components at a specific contaminated site investigation approach is illustrated in **Figure** 7. The site investigation will begin with a site reconnaissance and inspection. Then, representative intrusive judgmental sampling as the primary approach, field screening, borehole logging as per USCS, sample collection, and analysis. A variety of field testing methods are often used by field investigators to aid in the preliminary site assessment delineation program and to facilitate selection of samples. Soil gas surveys are frequently used in the field as a means of detecting the presence of volatile organics (VOCs) in the soil. Headspace vapor analysis, this field testing method is commonly used for assessing conditions of the soil samples during a drilling and sampling program. The last stage of the soil

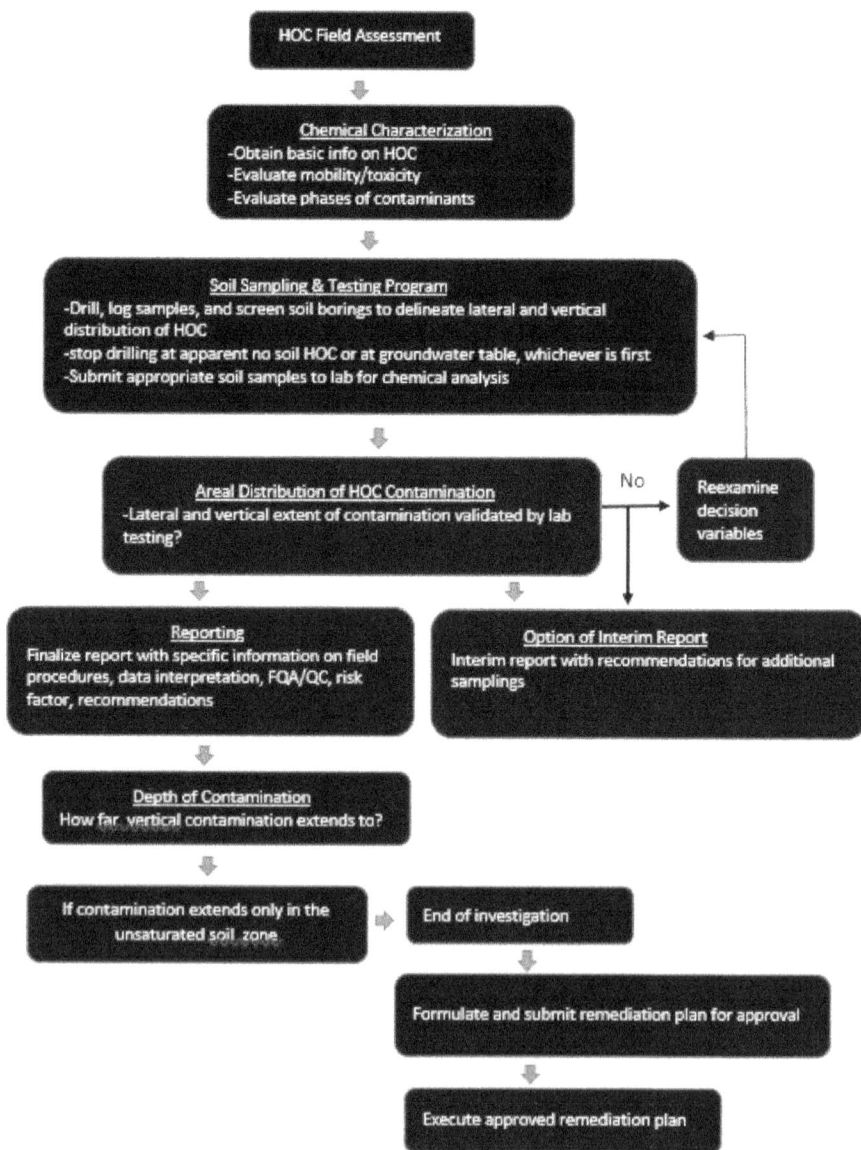

Figure 7.
General procedure of investigating HOC in soils.

investigation and surfactant-enhanced remediation program is the collection of confirmatory samples to determine whether or not the target clean-up goal has been achieved.

The mostly widely remedial methods for surfactant-enhanced remediation are in-situ flushing (washing), phytoremediation, ex-situ soil washing and ex situ bio-remediation. All methods require solubilization of the contaminant to be effective. However, in situ flushing solubilization must be accompanied with migration of the contaminant in the soil porous medium for collection, removal, and treatment. Each of these aforementioned methods is briefly discussed in the next sections.

7.1 In-situ flushing

In-situ soil flushing remediation method is a process that uses a flushing aqueous solution of surfactant to extract HOC by flooding the surface of a contaminated site or injection through vertical wells into a contaminated zone. Through continuous injection of the surfactant solution via the injection wells, contaminants partition into the flushing solution and leached into the soil. The mobilized contaminant-leaching solution flows through the contaminated zone and is extracted by downgradient extraction wells (**Figure 8**). The contaminant-flushing solution mixture is separated and treated or disposed of, or the treated effluent is reinjected. The physical and chemical properties of a soil, and the amount and type of surfactant solution are key factors in determining the efficiency of soil flushing [34]. However, some nonscientific factors including the cost of surfactant, dosage of surfactant solution, and the size of the contaminated site should be considered in order to ensure the economy of the remediation project.

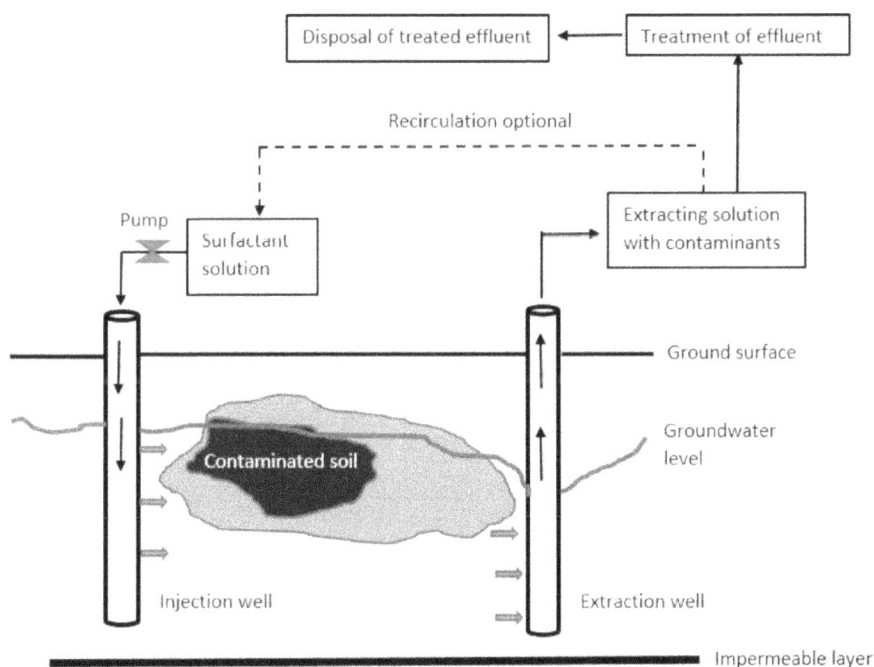

Figure 8.
Schematic of an in-situ flushing system for soil remediation.

7.2 In-situ phytoremediation

Phytoremediation remediation is a green technology technique that makes the use of plants as natural agents to absorb, degrade and/or sequester HOC over time in a contaminated soil. However, it can be slow and strategically should be used in a treatment train approach with in-situ flushing when feasible. Plants take up chemicals when their roots take dissolved chemicals and nutrients from the soil aqueous solution and additionally, HOC can be biodegraded by micro-organisms found in the plants rhizosphere. Efficacy of phytoremediation will depend on a combination multiple mechanisms in relation to specific plant species. The mechanisms may involve phytoremediation capacity, phytoaccumulation, phytovolatilization, rhyzodegradation, and phytodegradation. Crucial is optimizing surfactant-enhanced mass transfer of sorbed HOC in the aqueous soil solution particularly in the presence of multiple contaminants. Various contaminants may have different affinity for the soil sorbing sites which in return will affect the strength and mechanism of retention. When choosing plant species for a phytoremediation project, several relevant factors should be examined including type of plant root system, above ground biomass, depth of roots penetration, toxicity tolerance to the contaminants and surfactant, plant hardiness, depth of vertical contamination, adaptability to prevailing climatic conditions, resistance to diseases and pests, plant growth rate, nutrients requirement, and time required to achieve the desired level of cleanliness.

7.3 Ex-situ soil washing

Ex-situ soil washing is a mechanical process that involves delineating the areal extent of contamination, excavating the contaminated soil, pretreat it as necessary and then treat it with a surfactant solution. The soil washing can be performed in batch or continuous modes. The main steps are schematically depicted in **Figure 9**.

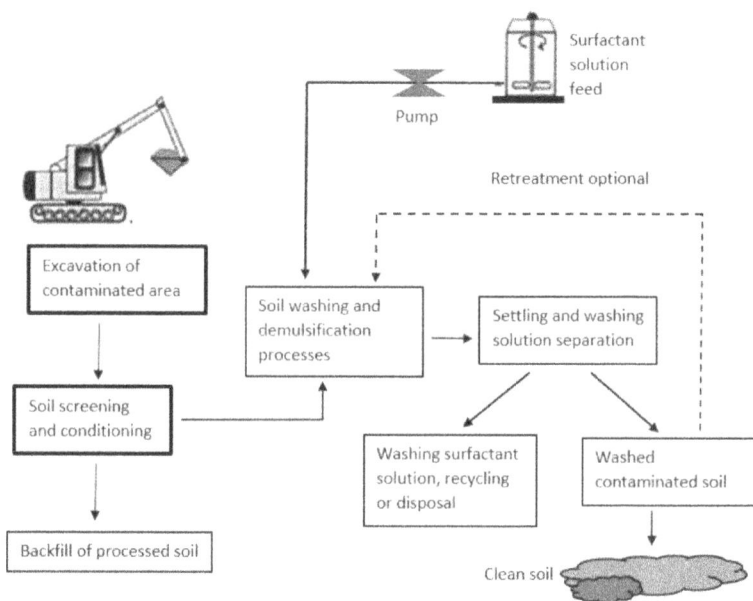

Figure 9.
Schematic diagram of ex-situ soil washing.

In practical term, ex-situ soil washing is considered a time-efficient and all-around technique, and a media transfer technology. It allows to treat a broad range of contaminant types and concentrations. Removal of coarse fractions is a key step and they can be reused on site as clean fill. However, there is a general held view that this technique is only cost efficient for coarse and granular soils where the clay and silt content make up less than 30% of the soil matrix. Factors that may limit the effectiveness and applicability of this method include effective removal of HOC sorbed onto clay-size particles by a surfactant, high soil humic content, and ambient temperature at treatment time.

7.4 Ex-situ surfactant enhanced bioremediation

Ex-situ surfactant enhanced bioremediation method refers to the biostimulation of soil natural biodegraders and increasing contaminant bioavailability. Two main prerequisites for biodegradation to take place are carbon source as electron donors and nutrients, as amendment. HOC contaminants in soils exhibit no or very low solubility at all and thermodynamically tend to partition to the soil solid phase. The concomitant effect is the level of hydrophobicity displays limits dissolved mass transfer phase and bioavailability, thereby limiting its biotic degradation in the soil system. Optimizing nitrogen and phosphorous status in the contaminated soil can have direct impact on contaminants biodegradation and microbial activity. This technique can be performed in various configurations which include windrow and various types of bioreactors. The general procedure of an ex-situ soil bioremediation is illustrated in **Figure 10**. Regardless of the system configuration and design emphasized, the treatment process must be optimized. Aqueous slurry conditions range from 20 to 40% w/v and should be not toxic to the soil microbial population. The slurry bioreactor sometimes may operate in sequencing batch reactors to achieve a desired treatment train objective (**Figure 10**). In this regard, dehalogenation conducted under anaerobic conditions is a prerequisite prior to

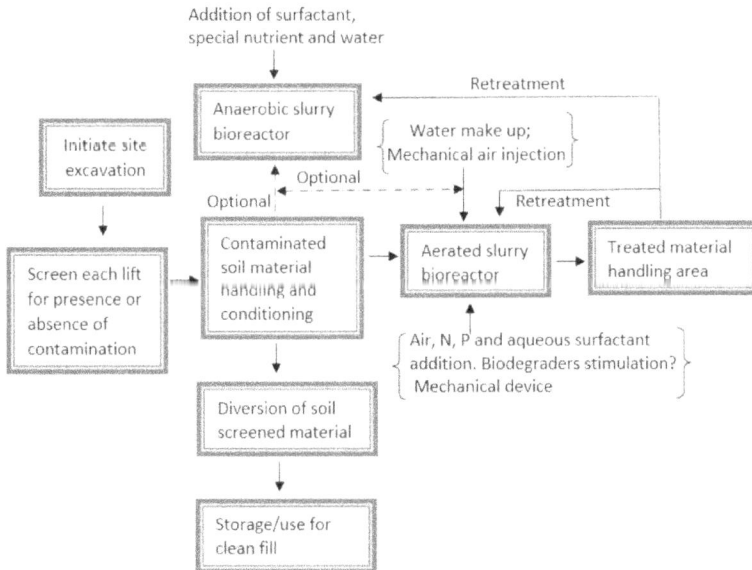

Figure 10.
Illustration of a typical batch sequencing slurry bioreactor (adapted with permission from [1]).

aerobic treatment. If dehalogenation is not required, the biodegradation treatment process can be performed under aerobic conditions only. Aerobism can be maintained during treatment by performing slurry mixing with mechanical or pneumatic devices in a rather intermittent than continuous mode. Mechanical mixing homogenizes the contaminant in the slurry bioreactor. A matrix summary of critical success factors for ex-situ surfactant enhanced bioremediation can be found elsewhere [1].

8. Mixing surfactants for their enhancement effect

Remediation of contaminated with mixed HOCs is generally very challenging and compounded due sorption on the soil matrix and different solubility properties. The strategy of mixing different classes of surfactants is to achieve a synergistic solubilization effect for the extracting solution. For example, when ionic and non-ionic surfactants are combined, the mixed surfactants solution results in a stronger solubilization effect than single surfactant solution. The reason is that nonionic surfactants diffuse the ionic surfactants and to some degree, reduce the influence of electrostatic repulsion between affecting the ionic surfactant molecules [35]. It has been reported that appropriate combination of several surfactants could inhibit the respective sorption of individual surfactant onto the soil. So, the loss of surfactant resulting from sorption is reduced and thereby increases the capability of mixed surfactants for HOC desorption in soils [36–39]. Synergistic effects of mixed surfactants in the binary blends can be best attributed to a decrease of CMC of the surfactant solutions, larger amount of available micelles formation, increase of MSR, lower polarity and higher aggregation of number of the mixture micelles.

9. The future of surfactants application for site clean-up

The potential adverse impact of HOC in soil has been a significant concern around the world for the public, policy makers, environmental regulators, and scientists. Even at very low concentrations and low solubility, these contaminants are generally considered highly toxic, mutagenic as well as carcinogenic, or can pose some other harm to humans and other ecological receptors. Costly site-specific remediation strategies have often been employed and too often with limited success. In many instances, site-specific remediation strategies are designed towards partial mass removal, plumes containment, source zone stabilization, relative to a formulated acceptable risk-management objective. The use of surfactants-aided soil remediation represents a technically attractive, cost-effective, and promising technology for reclaiming and rehabilitating contaminated sites. As a remediation technology, it is becoming well established because of its effectiveness and its promising results to retain the original nature of soil. Ideally, the primary goal of surfactant-aided remediation is to achieve 100% bioavailability and removal of contaminants with minimal xenobiotic effects and toxicity. Current research activities are very promising in this regard and continue to make more efficient synthetic and biosurfactants. However, there is an urgent need for both theoretical and empirical research on tertiary blends of surfactants-aided soil remediation and with additives mixed. More elaborative research works is also needed to elucidate the potential fate, characterization of soil and environmental interaction properties, health and ecological risks that may arise from surfactants entering the environment.

10. Conclusions

Surfactants-enhanced soil remediation represents an effective alternative to traditional remedial framework and has been successfully incorporated into various ex-situ and in-situ remediation technologies. There is a great potential to develop novel synthetic and biosurfactants that will exhibit higher biodegradability, less toxic, higher removal efficiency, more economical and more recyclable. Noteworthy are the prospects of the development and commercial production of mixed surfactants with low CMC containing additives mixed that will reduce remediation cost and increase remedial performance.

Author details

Roger Saint-Fort
Faculty of Science and Technology, Department of Environmental Science,
Mount Royal University, Calgary, Alberta, Canada

*Address all correspondence to: rsaintfort@mtroyal.ca

IntechOpen

References

[1] Saint-Fort R. Ex-situ surfactant enhanced bioremediation of NAPL impacted vadose zone. In: Larramendy, L.M., Soloneski, S., editors. Soil Contamination: Current Consequences and Further Solutions. InTech; December 2016. p. 307-327.

[2] Yang Y, Tao S, Zhang N, Zhang DY, Li XQ. The effect of soil organic matter on fate of polycyclic aromatic hydrocarbons in soil: a microcosm study. (2010) Environmental Pollution (5):1768-1774.

[3] Kohn N P, Evans N R. 2002. Phase I Source Investigation, Heckathorn Superfund Site, Richmond, California. PNNL-14088. Prepared for the U. S. Environmental Protection Agency by Battelle Marine Sciences Laboratory, Sequim, Washington; published by Pacific Northwest National Laboratory, Richland, Washington.

[4] Ashraf S, Ali Q, Zahir Z A, Ashraf S, Asghar H N. Phytoremediation: environmentally sustainable way for reclamation of heavy metal polluted soils. (2019) *Ecotox.* Environ. Safe. 174: 714–727.

[5] Song Y, Hou D, Zhang J, O'Connor D, Li G, Gu Q, Li S, Liu P. Environmental and socio-economic sustainability appraisal of contaminated land remediation strategies: A case study at a mega-site in China. (2018) Sci. Total Environ. 610-611:391-401.

[6] Rosen M J. Surfactants and Interfacial Phenomena, Wiley, New York, NY, 2nd Edition, 1989.

[7] Roy W R, Griffin R A. Surfactant-and-chelate-induced decontamination of soil materials: current status. Environmental Institute for Water Management Studies. Open File Report 21. The University of Alabama, AL, USA, 35487 (1988).

[8] Eljack M D, Hussam A. Extraction and solubilisation of crude oil and volatile petroleum hydrocarbons by purified humic and fulvic acids and sodium dodecylbenzenesulfonate. (2014) J. Environ. Sc. Health. A. 49: 1623-1630.

[9] Tang C F, Lian X J. Removal of diesel from soil using Rhamnolipid and sodium dodecyl sulfolane. (2013) Appl. Mech. Mater. 361-363:875-878.

[10] Peng S, Wu w, Chen J. Removal of PAHs with surfactant-enhanced soil washing: Influencing factorsand removal effitiveness. (2011) Chemosphere 82:1173-1177.

[11] Gitipour S, Narenjkar N, Farvash E S, Asghari H. Soil flushing of cresols contaminated soil: application of nonionic and ionic surfactants under different pH and concentrations. Journal of Environmental Health Science and Engineering 12:47.

[12] Liu F, Wang C, Liu X, Liang X, Wang Q. Effects of alkyl polyglucoside (APG) on phytoremediation of PAH-contaminated soilby an aquatic plant in the Yangtze Estuarine Wetland. (2013) Water, Air, Soil Pollut. 224:1633.

[13] Rosen M J. Geminis: A new Generation of Surfactants. (1993) Chemtech 23:30-33

[14] Menger F M, Littau C A. Gemini-surfactants: synthesis and properties. (1991) J. Am Chem. Soc. 113:1451.

[15] Villa R D, Trovo A G, Nogueira R F P. Soil remediation using a coupled process soil washing with surfactant followed by photo-Fenton oxidation. (2010) J. hazard Mater. 174:770-775.

[16] Zheng G, Selvam A, Wong J W C. Enhanced solubilisation and desorption of organochlorine pesticides (OCPs)

from soil by oil-swollen micelles formed with a nonionic surfactant. (2012) Environ. Sci. Technol. 46:12062-12068.

[17] Pei G, Zhu Y, Cai X, Shi W, Li H. Surfactant flushing remediation of o-dichlorobenzene and p-dichlorobenzene contaminated soil. (2017) Chemosphere 185:1112-1121.

[18] Ivey G A, Beaudoin M. Case study: In-situ surfactant enhanced DNAPL recovery pilot project, Refinery site. Montreal Canada. In: FCS 2010: Federal Contaminated Sites National workshop, Montreal, Quebec, Canada.

[19] *Thien* S J. A flow diagram for teaching texture by feel analysis. (*1979*) Journal of Agronomic Education 8: 54-55.

[20] Soil Texture, on the Web, [access online] http://www.fao.org/fishery/doc s/CDrom/ FAO_Training/FAO_Tra ining/General/x6706e/ x6706e06.htm. [accessed August 20, 2021].

[21] Zhentian S., Jiajun C., Jianfei L., Ning W., Zheng S. Anionic-nonionic mixed-surfactant-enhanced remediation of PAH-contaminated soil. (2015) Envron. Sci Pollut Res. 22(16): 12769-12774.

[22] Parekh P., Varade D., Parikh J., Bahadura P. Anionic-cationic mixed surfactant systems: micellar interaction of sodium dodecyl trioxyethylene sulfate with cationic Gemini surfactants. (2011) Colloids Surface A: Physicochem Eng Asp 385(1):111-120.

[23] Mohamed A., Mafoodh, A.M. Solubilisation of naphthalene and pyrene by sodium dodecylsulfate (SDS) and polyoxyethylenesorbitan monooleate (Tween 80) mixed micelles. (2006) Colloids Surface A: Physicochem Eng Asp 287(1):44-50.

[24] Mohamed A, Mafoodh A M. Solubilisation of naphthalene and

pyrene by socdium dodecylsulfate (SDS) and polyoxyethylenesorbitan monooleate (Tween 80) mixed micelles. (2006) Colloida and Surfaces A: Physicochemical and Engineering Aspects 287:44-50.

[25] Jackson M, Eadsforth C, Schowanek D, Delfosse T, Riddle A, Budgen N. Comprehensive review of several surfactants in marine environments: Fate and ecotoxicity. (2016) Environ Toxicol Chem. 35 (5): 1077-1086.

[26] Ying G G. Fate, behavior and effects of surfactants and their degradation products in the environment. (2006) Environ Int. 32 (3): 417-431.

[27] Cowan-Ellsberry C, Belanger S, Dorn P, Dyer S, McAvoy D, Sanderson H, Ersteeg D, Ferrer D, Stanton K. **Environmental Safety of the Use of Major Surfactant Classes in North America.** Critical reviews in Environmental Science and Technology. (2014) 44 (17): 1893-1993.

[28] Volkering F, Breure AM, Rulkens WH. Microbial aspects of surfactant use for biological soil remediation. (1997) Biodegradation 8: 401-417.

[29] Ishiguro L, Koopal, L K. Surfactants adsorption to soils component and soils. (2016) Adv Colloid Interface 231:59-102.

[30] Franzetti A, Di Gennaro P, Bevillacqua A, Papacchini M, Bestetti G. Environmental features of two commercial surfactants widely used in soil remediation. (2006) Chemosphere 62:1474-1480

[31] Bodarenko O, Rahman PKSM, Rahman TJ, Kharu A, Ivask A. Effects of rhamnolipids from *Pseudomonas aeruginosa* DS10-129 on luminescent bacteria: toxicity and modulation of cadmium bioavailability. (2010) Microb. Ecol.59:588-600.

[32] Yin G G, Kookana R S. Endocrine disruption: an Australian perspective. (2002) AWA J Water 29 (6):53-57.

[33] Deshpandle S, Shiau B J, Wade D, Sabatini D A, Harwell J H. Surfactant selection for enhancing ex situ soil washing. (1999) Water Res. 33:351-360.

[34] Zhou Q X, Sun F H, Liu R. Joint chemical flushing of soils contaminated with petroleum hydrocarbons. (2005) Environ Intl. 31:835-839.

[35] Lee J, Yang J S, Kim H J, Baek K, Yang J W. Simultaneous removal of organic and inorganic contaminants by micellar enhanced ultrafiltration with mixed surfactant. (2005) Desalination 184:395-407.

[36] Yang K, Zhu L Z, Xing B S. Enhanced soil washing of phenanthrene by mixed solutions of TX100 and SDBS. (2006) Environ. Sci. 40:4274-4280.

[37] Zhou W, Zhu L. Enhanced soil flushing of phenanthrene by anionic-nonionic mixed surfactants. (2008) Water Res. 42:101-108.

[38] Yu H, Zhu L, Zhou W. Enhanced desorption and biodegrdation of phenanthrene in soil-water systems with with the presence of anionic-nonionic mixed surfactants. (2007) J. Hazard. Mater. 142:354-36].

[39] Liu J, Chen W. Remediation of phenanthrene contaminated soils by nonionic–anionic surfactant washing coupled with activated carbon adsorption. (2015) Water Sci Technol 72 (9):1552-1560.